Power Management in Mobile Devices

Power Management in Mobile Devices

Findlay Shearer

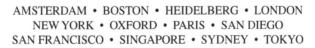

AMSTERDAM • BOSTON • HEIDELBERG • LONDON
NEW YORK • OXFORD • PARIS • SAN DIEGO
SAN FRANCISCO • SINGAPORE • SYDNEY • TOKYO

Newnes is an imprint of Elsevier

ELSEVIER

Newnes

Newnes is an imprint of Elsevier
30 Corporate Drive, Suite 400, Burlington, MA 01803, USA
Linacre House, Jordan Hill, Oxford OX2 8DP, UK

Recognizing the importance of preserving what has been written,
Elsevier prints its books on acid-free paper whenever possible.

Library of Congress Cataloging-in-Publication Data
Application submitted

British Library Cataloguing-in-Publication Data
A catalogue record for this book is available from the British Library.

ISBN: 978-0-7506-7958-9

For information on all Newnes publications visit our
Web site at http://www.books.elsevier.com

Transferred to Digital Printing, 2013

Printed and bound by CPI Group (UK) Ltd, Croydon, CR0 4YY

Working together to grow
libraries in developing countries

www.elsevier.com | www.bookaid.org | www.sabre.org

ELSEVIER   BOOK AID
International   Sabre Foundation

To Barbara, Amanda and Andrew
Special thanks to Andrew

Contents

Preface

Mobile wireless devices with monochrome screens that offer hundreds of hours of standby time and up to a working day of talk time are primarily voice devices and are now seen as dated. Consumers demand mobile devices with color screens, faster processors, and more storage, often with Wireless LAN, Bluetooth, Assisted Global Positioning System (AGPS), and broadband "always on" wireless connectivity. The power challenge is even more exacerbated with the advent of high-speed packet access (HSPA) at 10 Mbps. Applications such as games; AGPS and video conferencing make mobile wireless devices far more power hungry.

Batteries have evolved through sealed lead acid, nickel cadmium, nickel metal lithium ion, and lithium-ion polymer. The challenge is that the energy density supplied by the battery is not keeping up with the demand. Fuel cells offer a solution to the challenge. However, there are still a few years away from mass commercialization for mobile devices.

The management of energy consumption, for improved battery life, is widely considered to be the limiting factor in supporting the simultaneous needs of a low cost, high performance, feature rich mobile device in a small form factor. The current problem is defined in terms of technology gaps. Given that processor performance (Moore's law) doubles every 18 months, communications system performance (Shannon's Law) doubles every 8.5 months, and battery energy density only doubles every 10 years highlights a significant technology gap.

To bridge the supply/demand power gap, engineers and scientists from such diverse areas as physics, chemistry, mechanical engineering, electrical engineering, biology, and computer science, have developed an arsenal of technologies ranging from manufacturing processes, tools, software, hardware, and circuit innovations that combined will deliver the expected user experience.

The energy conservation discipline is alive and thriving. Driven by the popularity for mobile devices and subsequent R&D investments, scientists and engineers are continually developing creative and innovative solutions required for commercial success of mobile products. However, without a "crystal ball" only time will tell which solutions will ultimately succeed in solving the problem of efficient energy utilization.

Scope and Outline of the Book

The book provides an in-depth coverage of the technical challenges the mobile device industry has to embrace and resolve to meet the ever growing consumer demands for nomadicity and the insatiable demand for smaller form factor, lower cost, feature rich mobile devices with longer battery lives.

A pictorial representation of the contents of the book is shown in page xv.

Chapter 1, "Introduction to Power Management in Portable Personal Devices" discusses the growing trend for mobile devices, including smartphones, portable media players, game machines, and portable navigation devices. In addition, a deep dive of the most ubiquitous mobile device, the cellular phone, is presented. This includes cellular technology operation and evolution and how it is paired with Bluetooth and Wi-Fi to provide seamless mobility.

Chapter 2, "Hierarchical View of Energy Conservation", is concerned with the "technology gaps" that have widened over time due to the different rates that various components, that comprise a mobile device, have evolved. These gaps include the microprocessor and memory bandwidth gap; power reduction gap; and algorithmic complexity gap. A top-down holistic approach is required to address the technology gaps. Manufacturing processes, transistors, and packaging elements, of the holistic solution to the technology gaps, are presented in the chapter. Multi-gate transistors, copper interconnects and low-k dielectrics are some of the key technologies described. In addition, the benefits of packaging techniques, such as System-in-a-Package (SiP) and Package-on-Package (PoP), are presented.

Chapter 3, "Low Power Design Techniques, Design Methodology, and Tools", focuses on the role played by low power design techniques, such as dynamic process temperature compensation, static process compensation, power gating, state retention power gating, and clock gating and asynchronous techniques, for energy conservation. Low power System-on-a-Chip (SoC) design methodologies, tools and standards are also addressed. In the standards section, the Common Power Format and Unified Power Format are reviewed.

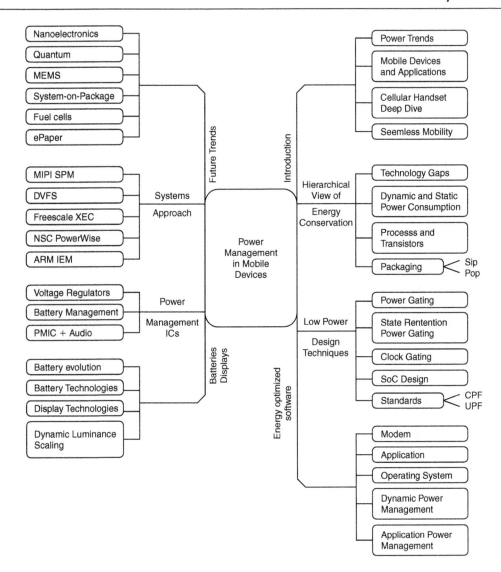

Chapter 4, "Energy Optimized Software", covers the software platforms and components that constitute a mobile device. Software mobile platforms, including Microsoft Windows Mobile, Symbian, Linux or RTOSs, as well as middleware runtime Java and Brew, are addressed. In addition, the chapter addresses the numerous software techniques employed to conserve energy. These techniques include dynamic power management, energy efficient compilers, and application-driven power management. The chapter concludes by highlighting the growing importance of software in today's mobile device.

Chapter 5, "Batteries and Displays for Mobile Devices" addresses two of the most important components today in a mobile device, the battery and the display. Battery evolution, from sealed lead acid to lithium-ion polymer, is reviewed. Also battery fundamentals and chemistry selection are addressed in depth. In addition, portable device display technologies, and their significant impact on energy consumption in a mobile device, are addressed. Display technology approaches, like emissive, transflective, reflective, and transmissive, are described. Key energy conserving LCD techniques, including dynamic luminance scaling and backlight auto regulation, are also covered in this chapter.

Chapter 6, "Power Management Integrated Circuits" focuses on how power management needs have proliferated exponentially with the variety of mobile devices, features and functions growing enormously in the recent years. As long as the mobile phone was simply required to make and receive phone calls, it embodied one set of power management requirements. However, as mobile phone manufacturers have heaped on the features, each requirement placed another demand on power management. Major power management components, such as linear (low dropout, LDO) and switching regulators (buck, boost), are covered in depth. Also battery management functions, such as fuel gauges, charging, authentication, and protection, are also considered in Chapter 6. The chapter concludes by describing next generation PMICs that go beyond simple signal conditioning and distribution of power. By integrating a wide range of functions tailored to specific applications, such as full-featured audio paths with analog, digital, and power audio interfaces, touch screen support, coin cell backup supply switching and charging, backlighting, LED drivers, and regulators optimized for specific functions such as cellular radios.

Chapter 7, "System Level Approach to Energy Conservation", addresses the need for an entire system approach to energy conservation. With a highly integrated systems approach, an important element, the Power Management IC, enables developers to optimize power consumption at the system level, while significantly reducing the design complexity required to achieve these gains. Companies such as AMD, Intel, Texas Instruments, Freescale Semiconductor, ARM, National Semiconductor, and Transmeta have obtained good results using different levels of a system approach to energy conservation. In addition to the proprietary commercial approaches, the Mobile Industry Processor Interface (MIPI) System Power Management (SPM) Architectural Framework is described.

Chapter 8, "Future Trends in Power Management", starts by presenting the ever demanding future requirements of mobile devices. Cellular download data rates of 100 Mbps and high definition video are some examples of power hungry technologies.

One conclusion reached in this chapter is the paramount role manufacturing processes will play in the future of energy conservation. From thin body to Fin-FETs, heterogeneous materials including high-k metal gates, Micro-Electrical and Mechanical Systems (MEMS), nanoelectronics, quantum computing, and genetic engineering will all contribute to achieve the goal of energy conservation. In addition, packaging technologies, like Freescale's Redistributed Chip Packaging and System-on-a-Package, will play a key role on the path to efficient energy utilization. Finally, the importance of fuel cell technology and ePaper displays in energy conservation, are highlighted in Chapter 8.

Target Audience

The target audience may be a diverse group covering all aspects of technology and product development from the mobile device industry and academia. This audience includes, but is not limited to technical managers, software developers, and hardware designers, manufacturing engineers, technical marketers, strategists, analysts, and business managers. The book should also appeal to students taking senior or graduate level mobile computing courses and those with an interest in working in the mobile device industry and its related value chain.

About the Author

The author, Findlay Shearer, holds a B.S.E.E. from Glasgow Caledonian University and a M.S.S.E. and M.B.A. from University of Texas at Austin. He is currently a Senior Product Manager in Freescale Semiconductor, Inc.

CHAPTER 1

Introduction to Power Management in Portable Personal Devices

The number of personal portable devices sold each year is increasing rapidly. Cell phones are ubiquitous. Worldwide sales for 2007 are shown in Figure 1.1[1]. The mobile phone industry is currently the largest consumer electronics (CE) segment in the world. The cell phone has replaced the personal computer as the most universal piece of technology in our lives. Industry analysts indicate that cellular phones are outpacing personal computers at the rate of 5 to 1.

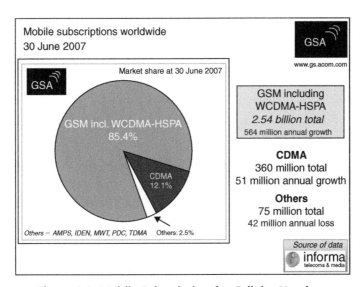

Figure 1.1: Mobile Subscription for Cellular Handsets

Personal mobile devices are increasingly becoming more than just devices for voice communication as they have a multitude of features including connectivity, enterprise, and multimedia capabilities.

Another growing application area for personal portable devices is entertainment. Devices like portable media players (PMP), FM radios, MP3 players, and portable gaming devices are found in every electronics store. Portable music has improved significantly since the days of the cassette tape; a collection with 500 h of music now fit inside a shirt pocket.

In addition, gaming devices provide portable entertainment. Portable gaming devices were pioneered by Nintendo with the Gameboy. The gaming devices evolved from simple toys into powerful computers with the progress of technology. These gaming devices turn kids into young consumers that think digitally and prepare them for the mobile device market. Figure 1.2 shows the Sony Play Station Portable which is capable of playing games, music, video, and has a Wi-Fi connection.

Figure 1.2: Portable Game Machine Sony PSP

Source: http://www.sony.com

Information access, manipulation, and processing are also significant markets for portable devices. Instead of using a paper calendar, people are turning to a mobile device to organizing their lives.

Analog cameras have been replaced with digital counterparts. Laptop computers are dropping in price and have become affordable for a large audience. In business, the laptop computer is common. Laptops with wireless connections to the Internet either via Wi-Fi or cellular networks are now appearing, enabling full user mobility and broadband speed equivalent to their wired competitors like cable or DSL. Access to the Internet from a laptop offers the user a rich set of services. However, information access from portable devices is still underdeveloped.

A significant trend in portable devices is to reduce size. For several functions, devices have been shrunk toward their ultimate size: the size of an ordinary wrist watch. An even smaller size would further compromise usability. Wrist watch models exist on the market for cellular, global positioning system (GPS) location, photo camera, and MP3 audio applications. An increasing number of devices become more versatile, programmable, and flexible. For example, many mobile phones can also play music, take pictures, and have location-based capabilities.

This versatility also stimulates the shift toward multi-modality. Multi-modality means that the user interface supports multiple methods of interaction with the device, such as touch screens, speech, motion, or even gestures for portable devices equipped with a camera. Portable devices for information are no longer limited to plain text and devices for communication provide more than mere voice services.

The future of portable devices is difficult to forecast. The future personal portable devices are more than just devices for voice communication. Sure they have voice capabilities; however, it can replace a laptop with 100 Mbps data rate, e-mail, Internet browsing, and e-commerce. In addition it has a full range of multimedia capabilities including 8Mpix camera, camcorder, HD video, TV capability, and security capability to protect high-value content like movies and software.

Within this future scenario, the dominant type of device will be the generic multi-purpose device, similar to the device in Figure 1.3, that can be used for a wide range of applications.

Figure 1.3: Converged Mobile Phone, i-mate JASJAR

Source: http://www.mymobilepc.com

However, there are severe problems with mobile devices. The cell phone has a limited talk time and may die in the middle of a conversation. The mobile MP3 music player can store a collection of 500 h, but the batteries last less than 24 h. The portable gaming device has a display that is very difficult to read. A laptop only works for a few hours; after that it just becomes an unusable brick. The GPS locator that's on your wrist can only measure your position continuously for 4 h on one battery. The above examples illustrate that battery lifetime is a shared problem for portable devices.

The fundamental components of a wireless multimedia device are shown in Figure 1.4. The wireless multimedia devices that have been created are still not up to the task. More research and development is needed. Common problems with such devices are their insufficient performance, large size, high price, and limited battery lifetime.

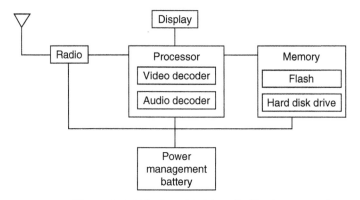

Figure 1.4: Wireless Multimedia Device

1.1 Power Trends

Power consumption is the limiting factor for the functionality offered by portable devices that operate on batteries [2]. This power consumption problem is caused by a number of factors. Users are demanding more functionality, more processing, longer battery lifetimes, and smaller form factor and with reduced costs.

Battery technology is only progressing slowly; the performance improves just a few percent each year. Mobile devices are also getting smaller and smaller, implying that the amount of space for batteries is also decreasing. Decreasing the size of a mobile device results in smaller batteries, and a need for less power consumption. Users do not accept a battery lifetime of less than 1 day; for personal portable devices even lifetimes of several months are expected.

New, more powerful processors appear on the market that can deliver the performance users desire for their new applications. Unfortunately, these powerful processors often have higher power consumption than their predecessors. Figure 1.5 graphically depicts the gap between battery energy density and the overall system performance required by evolving cellular and consumer portable devices.

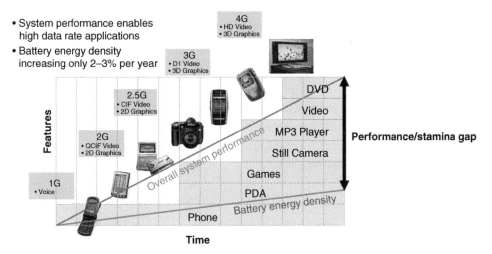

Figure 1.5: Performance/Stamina Gap [3]

Users want more features such as multimedia, mass storage, always-on wireless access, and speech recognition. It is important to utilize the available energy within batteries as efficiently as possible to meet user demands. Energy preservation, or energy management, is further translated into low power consumption by all parts of a portable device. Sophisticated power management is an important requirement to increase the battery life, usability, and functionality of mobile devices.

Power consumption is the rate at which energy is consumed. With the fixed energy capacity of a battery, the power consumption directly determines the lifetime of a portable device. The challenge is doing as much as possible with the lowest amount of energy. Efficiency is the key to solve the power crisis. High performance with high power consumption does not necessarily mean less energy efficient and conversely, low performance and low power consumption does not mean that a device is more energy efficient.

The power consumption of personal portable devices is not dominated by a single component, hardware or software. Several studies have investigated the power consumption of portable devices [4]. The main conclusion is that there is no single component or single activity that dominates the power consumption in a portable device. Therefore, the power consumption of all components needs to be reduced to lower the total amount of power.

Unfortunately, battery technology is not improving at the same pace as the energy requirements of handheld electronics. Therefore, energy management, once in the realm of desired features, has become an important design requirement and one of the greatest challenges in portable computing for a long time to come.

1.2 Mobile Devices and Applications

There has been an explosion of battery-powered mobile devices in the market. They range from cellular phones, PMP, portable audio players, portable navigation devices (PNDs), portable game machines, translators, cordless phones, remote controls, and digital cameras to name a few. Table 1.1 highlights the diversity of portable applications, user expectations, and system requirements.

Table 1.1: Diversity of Mobile Applications and User Expectations

Application Segment	User Expectations	Application Requirements
Cellular phones	Receive and make voice calls, send and receive photos and videos, listen to music, play games for extended period before charging the battery	Medium to high performance Long usage time Long standby time
Digital cameras	Take many photos, quickly review, transfer them to a mass storage device via wire line and wireless technologies	Moderate to high performance Long play time Standby irrelevant
Handheld gaming	Play games for long time before replacing batteries	High performance Long play time Standby irrelevant
Personal digital assistants	Do everything a laptop can for extended period before having to plug into the wall	High performance Long play time Standby irrelevant
Portable media players	Listen to music, watch movies for long time without interruption to recharge	Moderate to high performance Long play time Standby irrelevant

1.2.1 Cellular Phones

When the first cellular handsets were shipped more than 20 years ago, no one could have imagined all the components and technologies that would be squeezed into a handset 20 years later. Back then the DynaTAC 8000X handset from Motorola, fondly called the "Brick," was designed to carry on a voice call, and not much more than that. It weighed 2 pounds, offered half an hour talk time and cost almost $4,000.00 (Figure 1.6).

Figure 1.6: DynaTAC 8000X

Source: http://www.motorola.com

Today, there are some ultra low-cost handsets that just handle voice, but that is more an exception than the rule.

Cellular handsets vendors have to produce devices with an ever-increasing number of new technologies. Bluetooth, Wi-Fi, audio, video, cameras, TV, GPS, and others will continue to be blended into handsets in every combination imaginable, and the pressures to make these handsets as cheap as possible, in a "palm" size form factor and without sacrificing battery life, will only increase (Figure 1.7).

1.2.1.1 Cellular Phone Applications

Cellular phone applications can range from the sublime to the ridiculous. The most common application known by many as the killer application is the voice call. When all else fails on your cellular phone you at least want to be able to make a call.

Currently, there are a number of handsets commonly known as Smartphones or portable multimedia computers. Two key contenders in this market segment include the Nokia N95 and Apple iPhone (Figure 1.8).

Figure 1.7: Applications Converging on a Cellular Device

Source: http://www.freescale.com

Figure 1.8: Nokia N95

Source: http://www.nokia.com

The Nokia N95 delivers the following functionality:

- **Search**
  - Built-in GPS mapping
  - Web Browser with Mini Map
  - Location-based mobile searcher

- **Connectivity**
  - View e-mails with attachments
  - Play your videos on compatible home electronic devices via TV-out cable
  - Connect to your compatible PC via USB 2.0, Bluetooth wireless technology, or WLAN

- **MP3 player**
  - Stream your music wirelessly to your compatible headphones, home or car stereo
  - Hook up your favorite headphones or home stereo speakers
  - Store up to 2 GB of sound with expandable memory

- **Imaging**
  - 5 megapixel camera
  - Optics
  - 2.6" display
  - Shoot in DVD-like quality video up to 30 frames per second
  - Post your photos directly to flickr™

Today's top of the line handset, with every conceivable high-end feature, is a fully Internet-enabled phone with the following specifications.

1.2.1.2 General

Operating frequency

- WCDMA2100 (HSDPA), EGSM900, GSM850/1800/1900 MHz (EGPRS)
- Automatic switching between bands and modes

Dimensions

- Volume: 90 cc

- Weight: 120 g

- Length: 99 mm

- Width: 53 mm

- Thickness (max): 21 mm

Memory functions

- Up to 160 MB internal dynamic memory for messages, ringing tones, images, video clips, calendar notes, to-do list, and applications

- Memory card slot supporting up to 2 GB microSD memory cards

Power management

- Battery: 950 mAH

- Talk time: up to 160 min (WCDMA), up to 240 min (GSM)

- Standby time: up to 200 h (WCDMA), up to 225 h (GSM)

Displays

- Large 2.6" QVGA (240×320 pixels) thin-film transistor (TFT) display with ambient light detector and up to 16 million colors

User interface

- Open operating system: Windows Mobile, Symbian, Linux

- User interface

- Dedicated Media Keys

- Multimedia Menu

- Active standby screen

Call management

- Contacts: advanced contacts database with support for multiple phone and e-mail details per entry, also supports thumbnail pictures and groups

- Speed dialing

- Logs: keeps lists of your dialed, received, and missed calls

- Automatic redial

- Automatic answer (works with compatible headset or car kit only)

- Supports fixed dialing number, which allows calls only to predefined numbers

- Push to talk over cellular (PoC)

- Conference call

Voice features

- Speaker-independent name dialing

- Voice commands

- Voice recorder

- Talking ring tone

- Integrated hands-free speaker

Messaging

- Text messaging: supports concatenated SMS, picture messaging, and SMS distribution list

- Multimedia messaging: combine image, video, text, and audio clip and send as MMS to a compatible phone or PC; use MMS to tell your story with a multi-slide presentation

- Automatic resizing of your megapixel images to fit MMS

- Predictive text input: support for all major languages in Europe and Asia-Pacific

Data transfer

- WCDMA 2100 (HSDPA) with simultaneous voice and packet data (PS max speed UL/DL = 384/3.6 MB, CS max speed 64 kbps)

- Dual Transfer Mode (DTM) support for simultaneous voice and packet data connection in GSM/EDGE networks. Simple class A, multi-slot class 11, max speed DL/UL = 177.6/118.4 kbps

- EGPRS class B, multi-slot class 32, max speed DL/UL = 296/177.6 kbps

1.2.1.3 Imaging

Imaging and video

- Up to 5 megapixel (2592×1944 pixels) camera, high-quality optical lens, MPEG-4 VGA video capture of up to 30 fps

- Direct connection to compatible TV via Nokia Video Connectivity Cable (CA-75U, included in box) or wireless LAN/UPnP

- Front camera, CIF (352×288) sensor

- Video call and video sharing support (WCDMA network services)

- Integrated flash

- Digital stereo microphone

- Flash modes: on, off, automatic, redeye reduction

- Rotating gallery

- Online album/blog: photo/video uploading from gallery

- Video and still image editors

- Movie director for automated video production

Mobile video

- Video resolutions: up to VGA (640×480) at 30 fps

- Audio recording: AAC mono

- Digital video stabilization

- Video clip length: limited by available memory

- Video file format .mp4 (default), .3gp (for MMS)

- White balance: automatic, sunny, cloudy, incandescent, fluorescent

- Scene: automatic, night

- Color tones: normal, sepia, black & white, negative, vivid

- Zoom: digital up to 10× (VGA up to 4×)

Mobile photography

- Image resolution: up to 5 megapixel (2592×1944 pixels)

- Still image file format: JPEG/EXIF

- Auto focus

- Auto exposure – center weighted

- Exposure compensation: $+2 \sim -2\text{EV}$ at 0.5 step

- White balance: automatic, sunny, cloudy, incandescent, fluorescent

- Scene: automatic, user, close-up, portrait, landscape, sports, night, night portrait

- Color tone: normal, sepia, black & white, negative, vivid

- Zoom: digital up to 20× (5 megapixel up to 6×)

Camera specifications

- Sensor: CMOS, 5 megapixel (2592×1944)

- High-quality optical lens

- Focal length 5.6 mm

- Focus range 10 cm~infinity

- Macro focus distance 10–50 cm

- Shutter speed: mechanical shutter: 1/1000~1/3 s

1.2.1.4 Music

Music features

- Digital music player – supports MP3/AAC/AAC+/eAAC+/WMA/M4A with playlists and equalizer

- Integrated hands-free speaker

- OMA DRM 2.0 & WMDRM support for music

- Stereo FM radio (87.5–108 MHz/76–90 MHz)

Radio

Listen to music and interact with your favorite radio stations

- Find out what song is playing, who sings it, and other artist information

- Enter contests and answer surveys, vote for your favorite songs

1.2.1.5 Explore

Navigation

- Built-in GPS

1.2.1.6 E-mail

- Easy-to-use e-mail client with attachment support for images, videos, music, and documents

- Compatible with Wireless Keyboard

1.2.1.7 Browsing

- Web Browser with Mini Map

1.2.1.8 Digital Home

- Play video, music, and photos on home media network – compatible TV, stereo, and PC over WLAN/UPnP

1.2.1.9 Java Applications

- Java MIDP 2.0, CLDC 1.1 (Connected Limited Device Configuration (J2ME))

- Over-the-air download of Java-based applications and games

1.2.1.10 Other Applications

- Personal information management (PIM)

- Advanced PIM features including calendar, contacts, to-do list, and PIM printing

- Settings Wizard for easy configuration of e-mail, push to talk, and video sharing

- Data transfer application for transfer of PIM information from other compatible wireless devices

- WLAN Wizard

1.2.1.11 Connectivity

- Integrated wireless LAN (802.11 b/g) and UPnP (Universal Plug and Play)

- Bluetooth wireless technology with A2DP stereo audio

- USB 2.0 via Mini USB interface and mass storage class support to support drag and drop functionality

- 3.5 mm stereo headphone plug and TV-out support (PAL/NTSC)

- PC Suite connectivity with USB, Infrared, and Bluetooth wireless technology

- Local synchronization of contacts and calendar to a compatible PC using compatible connection

- Remote over-the-air synchronization

- Send and receive images, video clips, graphics, and business cards via Bluetooth wireless technology

1.2.1.12 Video

RealPlayer media player

- Full-screen video playback to view downloaded, streamed, or recorded video clips

- Supported video formats: MPEG-4, H.264/AVC, H.263/3GPP, RealVideo 8/9/10

High data rates, multimedia, and peripheral functions drive higher power requirements.

1.2.2 Portable Media Players

PMP are handheld devices whose primary capability is video playback. For the large majority of these products, the color screen is 3.5-in. or larger. Typically, PMPs also play music and display still images stored on embedded flash memory or hard disk drives (HDD). Some PMPs are based on Microsoft's Portable Media Center (PMC) platform and include products from iRiver, Creative, and Samsung (Figure 1.9).

PMPs that are available today can play Windows Media Audio (WMA) and MP3 audio, as well as Windows Media Video (WMV), MPEG-4, and other compressed video formats. Many of these players connect directly to PCs so that users can transfer content from the computer to the PMP, making the content portable. Other players offer the

Figure 1.9: Clix Portable Media Player

Source: http://www.iriver.com

ability to act as a portable Personal Video Recorder (PVR), recording directly from the TV to the device.

The functions of PMPs primarily revolve around video, audio, and images. Most devices will support a variety of data formats, although some products offer varying levels of native codec support. Some of the functions available on these products include:

- Transfer images from camera using USB 2.0 or USB On-The-Go (OTG)

- View images on LCD or on TV from device

- Create picture slideshow with or without music soundtrack

- View, store, edit, and create slideshows on device without a PC

- Playback and editing of video on device

- Download from camcorder to device

- Content from TV (PVR-like) downloaded to device from PC (TV tuner installed) or from a TV or other Set Top Box (STB)

- Video download sites accessed via PC or the device itself

A block diagram of a typical PMP is shown in Figure 1.10.

1.2.3 Portable Digital Audio Players

Portable digital music players, otherwise known as the MP3 players, are handheld devices that operate by compressing music files onto various memory media. Most portable digital music players contain similar components; but integration remains key as silicon manufacturers strive to address power consumption issues.

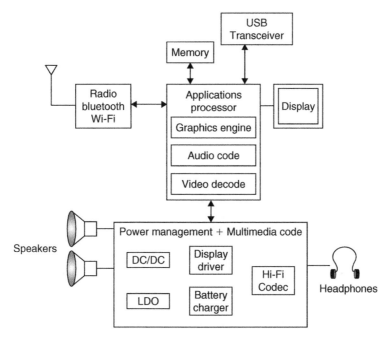

Figure 1.10: Block Diagram of a PMP

One of the key components in a portable digital audio player is the Applications Processor. Battery life continues to be an important issue in this device. Processor integration continues, as manufacturers begin to incorporate faster transfer technologies and support for more multi-memory functionality (Figure 1.11).

Figure 1.11: Sandisk Sansa Portable Digital Audio Player

Source: http://www.sandisk.com

The majority of MP3 player manufacturers are offering product lines that use two different memory formats: HDD and Flash memory. Flash-based, digital audio players are shockproof, dustproof, immune to magnetic fields, and very small. While some products have embedded Flash memory, some also have expansion slots to accommodate extra memory. The leading Flash card formats include: CompactFlash, SmartMemory, MultiMemory Card (MMC), Secure Digital (SD), and Memory Stick. HDD products offer users high capacity, portability, and versatility. Another advantage the HDD products have is they are capable of storing multiple data types.

Flash memory will remain the larger segment in memory technology this year because prices are falling rapidly and densities are increasing. In addition, Flash memory is smaller and consumes less battery power. We expect that Flash memory will be the primary memory used in MP3 players that are priced aggressively and that require low power consumption.

1.2.4 Portable Navigation Devices

GPS technology has been available for decades. The technology was initially used for aviation and government purposes. As GPS costs continue to fall, the technology is now being integrated into a number of today's portable Consumer Electronics products including Portable Navigation Devices (PND), Smartphones, and GPS-enabled PDAs. An example of a Garmin PNDs is shown in Figure 1.12.

Figure 1.12: PND, Garmin

Source: http://www.garmin.com

PNDs autonomously connect with a GPS and acquire radio signals to determine their location, speed, and direction. Autonomous GPS was initially used by the military and transportation industry and is a free public good for civilian use.

The standard GPS solution requires an analog GPS receiver, digital baseband, application processors, and memory semiconductor components (see Figure 1.13). The GPS receiver receives the GPS signal; the GPS baseband processor decodes the data and then sends the positioning information to the application processor in NMEA protocol or binary code; and the application processor renders the data graphically on the PND display, along with other multimedia files.

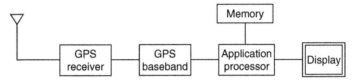

Figure 1.13: GPS Module Block Diagram

Software GPS (SW GPS) is an alternative GPS implementation that eliminates the need for a GPS baseband processor. Instead, a high performance application processor is enhanced with software to handle the intensive calculations. The SW GPS implementation is low cost, but the cost savings may be offset with the higher price of the required, high performance application processor (Figure 1.14).

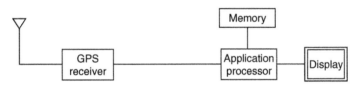

Figure 1.14: SW GPS Block Diagram

GPS technology was developed based on the assumption that the user would have access to the "blue-sky" to gather the satellite data. However, PNDs must also function when GPS signals are not available, such as within indoor environments (e.g. buildings or tunnels). Dead reckoning is a method used to enable PND navigation functionality when GPS signals are not available.

Dead reckoning uses sensors to calculate the location, speed, and direction of travel. There are two types of sensors available, Gyros and G-sensors. Gyros are very accurate in advancing a known position without a GPS signal. However, they are too expensive to be used in PNDs due to their size. Therefore, G-sensors are typically used to enable dead-reckoning functionality in PNDs.

In addition to Autonomous GPS, there are other enhanced GPS technologies that may be used in PNDs: Wide Area Augmentation System, Differential GPS, and Assisted GPS.

Assisted GPS (A-GPS) is commonly associated with location-based services (LBS) and uses an assistance server (Mobile Location Server) that connects to the PND through a cellular communication connection (i.e. embedded cellular module). The A-GPS Mobile Location Service has significant processing power and access to the reference network to enhance the positioning accuracy and can communicate with the PND in non-line of sight or indoor environments.

PND functionality is enhanced with a variety of multimedia technologies, which enable media-rich audio, speech, video, image, and graphics features.

1.3 Cellular Handsets: Deeper Dive

The most ubiquitous portable device is the cellular handset.

1.3.1 Cellular System Overview

In a cellular system the handsets carried by users are called Mobile Stations (MS). The Mobile Stations communicate to the Base Stations (BS) through a pair of frequency channels, one for uplink and another for downlink. All the BS of cellular systems are controlled by a central switching station called the Mobile Switching Center (MSC). The MSC is responsible for network management functions such as channel allocations, handoffs, billing, and power control.

The MSC is also connected to the Public Subscriber Telephone Network (PSTN) or Public Land Mobile Network (PLMN) enabling the MS to talk to a Land-Line telephone or vice versa.

The range of frequencies allocated to Mobile Cellular is divided into channels permitting many MS to talk simultaneously. This technique of multiplexing several users over a given spectrum is called Frequency Division Multiple Access (FDMA). Other techniques like Time Division Multiple Access (TDMA) and Code Division Multiple Access (CDMA) are also employed by modern cellular systems.

To support a large number of users in a given area a technique called "Frequency Reuse" is used by cellular systems. The available channels are divided into smaller groups. These groups of channels are allocated to a small region called "cell" which is hexagonal in shape. A group of cells is called a "cluster." Typical cluster sizes are 4, 7, 9, 12, and 19. Figure 1.15 shows a typical cellular system and cluster.

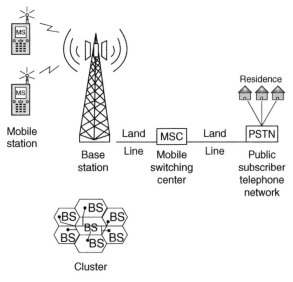

Figure 1.15: Cellular System

Another important feature of Mobile Cellular System is "Handoff." When an MS goes from one cell to another it needs to change to the BS of current cell site and possibly assigned a different set of frequency channels. This phenomenon is called "handover" or "handoff" (HO). The HO decision may be taken entirely by controlling station or may be coordinated by MS.

1.3.2 Evolution of Cellular Systems

A common way of classifying mobile phone technologies is to define them by generation. The first generation (1G) of mobile phones was based on analog cellular technology and the second generation (2G) is based on digital technology. The third generation (3G) provides digital bandwidth to support high data rates up to 2 Mbps. Technologies to upgrade 2G networks, but with no attempt to meet the performance of 3G networks, are often referred to as 2.5G. Figure 1.16 shows the evolution of the wireless generations and their data rates.

1.3.2.1 First Generation Analog Cellular

The introduction of cellular systems in the early 1980s represented a quantum leap in mobile communication with respect to capacity and mobility. Semiconductor technology and microprocessors made smaller, lighter weight, and more sophisticated mobile systems a practical reality for many more users. 1G cellular systems transmit only analog

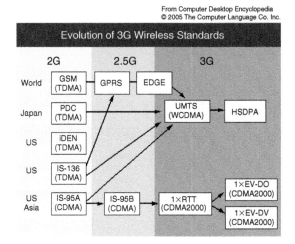

Figure 1.16: Evolution of Cellular

Source: Computer Desktop Encyclopedia. The Computer Language Co. Inc. 2005

voice information. The most prominent 1G systems are Advanced Mobile Phone System (AMPS), Nordic Mobile Telephone (NMT), and Total Access Communication System (TACS).

1G systems suffered from limitations of low service quality, long call setup time, inefficient use of bandwidth, susceptible to interference, bulky and expensive equipment, support only for speech and insecure transmissions.

1.3.2.2 Second Generation Digital Cellular

The development of 2G cellular systems in the 1990s was driven by the need to improve transmission quality, system capacity, and coverage. Further advances in semiconductor technology and microwave devices brought digital transmission to mobile communications. Speech transmission still dominates the airwaves. However, the demands for short text messages and data transmissions are growing rapidly. Supplementary services such as fraud prevention and encrypting of user data have become standard features that are comparable to those in fixed networks. 2G cellular handsets are classified based on the technique which is used to multiplex multiple users on to a single carrier frequency. They include Global System for Mobile Communication (GSM), Digital AMPS (D-AMPS), Code Division Multiple Access (CDMA), and Personal Digital Communication (PDC).

1.3.2.3 2G to 3G: GSM Evolution (2.5G)

At the end of the 20th century the transition form 1G to 2G was largely completed. The biggest trend was the move from voice-centric to data-centric systems. Phase 1 of the standardization of GSM900 was completed by the European Standards body ETSI in 1990 and included all necessary definitions for GSM network operations. Several services were defined including data transmission up to 9.6 kbps. GSM standards were enhanced in Phase 2 (1995) to incorporate a large variety of supplementary services that were comparable to digital fixed network ISDN standards. In 1996 ETSI decided to further enhance GSM in annual Phase 2+ releases that incorporate 3G capabilities. GSM Phase 2+ releases have introduced important 3G features such as Intelligent Network services; enhanced speech codecs enhanced full rate (EFR) and adaptive multi-rate (AMR), high data rate services and new transmission principles with High-Speed Circuit Switch Data (HSCSD), General Packet Radio Services (GPRS) and Enhanced Data rates for GSM Evolution (EDGE). Universal Mobile Telecommunications Standard (UMTS) is a 3G GSM successor standard that is downward compatible with GSM, using the GSM Phase 2+ enhanced Core Network.

1.3.2.4 Third Generation IMT-2000

3G standards were developed specifically to support high-bandwidth services.

The main characteristics of 3G systems, known collectively as IMT-2000, are a single family of compatible standards that are:

- Used worldwide

- Used for all mobile applications

- Support both packet and circuit switched data transmission that offer high data rates up to 2 Mbps (depending on mobility/velocity)

- Offer high spectrum efficiency

UMTS was developed by Third Generation Partnership Project (3GPP), a joint venture of several Standards Development Organizations. To reach global acceptance, 3GPP is introducing UMTS in Phases and Annual Releases. The first release (UMTS Rel. '99), introduced in December of 1999, defines enhancements and transitions for existing GSM networks.

The most significant change in release '99 was the new UMTS Terrestrial Radio Access (UTRA), a WCDMA radio interface for land-based communications. UTRA supports

Time Division Duplex (TDD) and Frequency Division Duplex (FDD). The TDD mode is optimized for public Micro and Pico cells and unlicensed cordless applications. The FDD mode is optimized for wide area coverage including public Macro and Micro cells. Both modes offer flexible and dynamic data rates up to 2 Mbps. Another newly defined UTRA mode, Multi Carrier (MC), is expected to establish compatibility between UMTS (WCDMA) and cdma2000.

1.3.2.5 Fourth Generation LTE, WiMAX, OFDMA, MIMO, etc.

Even though 3G still is in the early stages of ramp, work already has begun on the next generation called Long-Term Evolution (LTE). The primary difference between LTE and 3G is LTE's broadband wireless technology. LTE will be capable of supporting data rates of 100 Mbps downlink speed and 50 Mbps on the uplink and will enable your mobile device to transfer data as quickly as your computer can today when connected to a 100 base-T Ethernet line.

For comparison, 3G aspires to HSDPA (High-Speed Downlink Packet Access) 14.4 Mbps downlink and HSUPA (High-Speed Uplink Packet Access) 5.76 Mbps uplink speed, but presently only supports the downlink at 3.6 Mbps. Furthermore, 2G data rates are an order of magnitude lower than what 3G actually supports today.

Aside from high data throughput, LTE offers spectral efficiency improvement of three to four times compared with 3G's HSDPA. This allows for more users in the same bandwidth.

LTE also has scalable channel bandwidths from 1.25 to 20 MHz. With the flexibility that the scalable channel bandwidths provide, operators can optimize service to meet users' needs. In addition, LTE provides a 10 times reduction in latency to <10 ms. It is an entirely Internet protocol (IP)-based implementation with the design optimized to support mobility. These capabilities enable high-speed video and music downloads, interactive gaming and video broadcasting.

LTE technology is based on a couple of key concepts: Orthogonal Frequency Division Multiplexing (OFDM) and Multiple Input Multiple Output (MIMO). Traditional cellular systems exploit the two dimensions of frequency and time. MIMO technology introduces a third degree (space) of freedom that is one key to increase the data rate to 100 Mbps. Neither MIMO nor OFDM are new technologies. However, each requires a substantial amount of computational power. With smaller technology nodes combined with clever system and circuit design, implementation of these concepts is possible.

LTE's computational requirement is approximately 3,000 times that of basic 3G service. The peak transmit power requirement is twice that of basic 3G. Additionally, two more radio frequency (RF) bands need to be supported. At the same time, we must find implementation techniques that enable LTE-based phones at roughly the same size, price, and power consumption of today's 3G phones. Using traditional approaches will not satisfy these requirements.

Fourth generation (4G) cellular handsets will support TV in real time as well as video downloads at broadband speeds. They will also embrace automatic roaming to non-cellular systems such as Wi-Fi, satellite, and other wireless networks based on the users' requirements.

1.3.3 Cellular Handset Teardown

The components in wireless handsets can vary greatly by model, especially when additional multimedia features are added, but the primary functions are essentially the same. Voice traffic is captured by the antenna and processed by the RF block, which includes a power amp, an RF, and an IF (intermediate frequency) transceiver, which converts the analog signal to a digital bit stream. A digital baseband processes the encoded signal into voice signals and sends it through an audio DSP to the speaker. On the transmit side, the process is reversed: the microphone sends audio signals to the audio DSP, which are processed by the baseband and sent through the RF block to the antenna. In addition to these devices, an application processor or microcontroller controls the system and runs the phone software; this can be integrated into the digital baseband. The system also uses significant flash memory, SRAM, and power management devices (Figure 1.17).

1.3.3.1 Radio

Power amp: The power amp sits in front of the antenna and boosts signal strength.

Radio frequency: The RF front end controls the radio wave generation on the transmit side and receives the radio wave signal on the receive side, translating between low-frequency analog signals used in the handset (specifically by the IF transceiver) and high-frequency RF signals used in radio communications.

IF/transceiver/synthesizer: The IF performs all quadrature modulation of "I" and "Q" baseband signals, converting the analog radio signal to a digital stream to be processed by the baseband and vice versa.

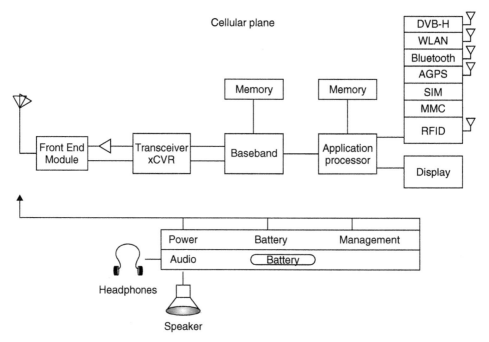

Figure 1.17: Wireless Handsets Anatomy

1.3.3.2 Baseband and Audio Processing

Digital baseband: The baseband processor performs digital processing on the signal, implementing the wireless modulation scheme (the encoding technique, e.g., GSM, CDMA, TDMA, GPRS, etc.), encoding and extracting voice signals in and out of the transmission protocol.

Audio DSP/baseband: Also called the analog baseband, the audio DSP sits between the digital baseband and the speaker/microphone circuits and processes the audio signals to and from the digital baseband, converting the stream into voice signals and vice versa.

Audio codec: On the receive side, the audio codec converts the digital voice signals to analog signals to be sent to the speaker amp; on the transmit side, it receives and converts voice signals from microphone to digital for processing by the audio codec.

Speaker amp: The speaker amp drives the audio signals in the handset speaker.

Depending on the architectural partitioning the audio codec and speaker amplifier can be integrated into the power management integrated circuit (PMIC).

1.3.3.3 Application Processor, Memory, and Power Management

Application processor/microcontroller: The application processor or microcontroller runs the system software and controls overall handset functionality. Most handsets integrate the processor into the digital baseband. More advanced handsets use a discrete processor.

Multimedia ICs: These processors are used in advanced phones to implement multimedia and connectivity functions. These features will usually be implemented with a chipset that includes digital processors plus an analog front end as described below. Note that some of this logic can be loaded onto the application processor, eliminating the need for discrete ICs.

Camera: Camera modules usually include a CMOS or CCD image sensor and a digital image processing IC.

Advanced graphics: Some handsets include a graphics co-processor IC, though many application processors can also perform this function.

Global positioning systems: GPS functionality requires an additional radio and baseband.

Bluetooth: Bluetooth functionality requires an additional radio and baseband.

Wireless LAN (Wi-Fi): 802.11 functionality requires an additional radio and baseband.

TV tuning: It requires a digital TV tuner and image processor.

MP3 audio decoding: MP3 audio decode can be done with a separate digital audio processor but can also be performed by an application processor.

MPEG-4 video processing: MPEG-4 video decode and encode can be done with a separate digital video processor but can also be performed by an application processor.

Flash memory: It is the primary memory used by the handset for storage of system software, phone book entries, and other user data. Nearly 100% of handset flash is high-density NOR flash, though NAND is making inroads as well, especially in multimedia handsets that require additional data storage.

SRAM: It is used by the digital baseband for digital processing and by the application processor. This can be standard SRAM or PSRAM (Pseudo-SRAM).

Stacked memory: It is a module containing flash memory and SRAM in a single package. Also called MCP (multi-chip package) memory. It is also sometimes offered with an application processor in-package. Additionally called Package-on-Package and System-in-Package.

Power management: Power management ICs regulate, control, manipulate, and optimize the use and delivery of battery power to handset components. In addition they can provide battery management capabilities including charging and protection.

1.3.4 Seamless Mobility: Connectivity

Seamless mobility, through additional wireless technologies, is key for mobile devices. A number of wireless technologies are integrated with cellular technologies to provide the user significant performance and cost options in service and application delivery.

Today's modern mobile device represents a convergence of communications and multimedia technologies. The proliferation of radios in modern mobile devices is remarkable. They include:

- Cellular including GPRS, EDGE, UMTS, HSDPA, HSUPA, IS-95, IS-2000, 1xEV-DO, and UMB

- Wireless MAN commonly known as mobile WiMax

- Wireless LAN defined by IEEE 802.11b, g, a, n. Commonly known as Wi-Fi

- Wireless PAN including Bluetooth and a migration path to Ultra Wide Band for high speeds

- Navigation including GPSs

- Broadcast mobile Digital TV (DTV) including Digital Video Broadcast – Handheld (DVB-H), Integrated Services Digital Broadcast – Terrestrial (ISDB-T), Digital Multimedia Broadcast (DMB), and Forward Link Only (FLO)

- Broadcast radio including FM, HD Radio, and DAB (Digital Audio Broadcast)

In addition to cellular radios the most common additional radios include Bluetooth and Wireless LAN.

1.3.4.1 Bluetooth

In many cases, portable devices do not have compatible data communication interfaces, or the interface requires cable connections and configuration procedures. A possible solution is to get rid of the cables and use short-range wireless links to facilitate on demand connectivity among devices. To allow pervasive adoption, the solution would also be low cost, enable compelling applications, and ubiquitously adopted by device suppliers.

One technology that has gained significant traction in the portable industry is Bluetooth. Bluetooth is a radio technology that promises to provide simultaneous wireless voice and data connectivity between notebooks, handheld data devices, and mobile phones, as well as providing a short-range wireless link to the Internet. Initially designed as a cable replacement technology, Bluetooth is a hardware/software module and is considered a Personal Area Network [5].

Bluetooth Technology Overview

Bluetooth is an open, "always-on" radio technology that enables low-power, short-range radio communication in the unlicensed ISM 2.4 GHz band. The low transmitting power of Bluetooth makes it ideal for small, battery-powered mobile devices such as PDAs, mobile phones, and portable PCs. Key technical specifications are outlined in Table 1.2.

Table 1.2: Key Technical Specifications

Normal range	30 ft (0 dBm)
Optional range	300 ft (+20 dBm)
Normal transmitting power	0 dBm (1 mW)
Optional transmitting power	−30 to +20 dBm (100 mW)
Receiver sensitivity	−70 dBm
Frequency band	2.4 GHz
Gross data rate	1 Mbps
Maximum data transfer	721 kbps
Power consumption: Hold/park	50 μA
Power consumption: Standby	300 μA
1.2.1 Power consumption: Max	30 mA
Power modules	Low-short range and high-medium
Packet switching protocols	Frequency hopping scheme
Topology	Flexible multi-Piconet structure

Bluetooth Components

Every Bluetooth system consists of a radio unit, a baseband unit (hardware), a software stack, and application software [6]:

- The radio unit or transceiver creates the air interface among devices located between 30 and 300 ft.

- The baseband unit is the actual enabler of the RF link. The Link Manager, software embedded in the baseband, discovers the remote Link Manager, requests the name of the device, gets its address, and establishes and controls the connection.

- The software stack enables application software to interface with the baseband unit.

- The application software defines the functionality of the device.

Bluetooth Applications

Initial applications for the technology include file transfers, data access points, file and data synchronization between devices, and wireless headsets for mobile phones and computers. Key applications for Bluetooth include:

- *Hidden computing*: The ability for a Bluetooth-enabled device to access and synchronize data. E-mail from a notebook could be displayed on a Bluetooth-enabled phone or handheld PC. With a Bluetooth-based wireless headset, a user could request a phone number from their handheld and then direct the mobile phone to dial the number.

- *"Pico" networks*: Ad hoc data exchanges (files, business cards, messaging, printing) based on establishing "trusted" relationships between devices and users. In this scenario, as many as eight notebooks and their users could be working collaboratively in a conference room.

- *Synchronization*: Unconscious synchronization between a user's multiple devices. For example, a user walks into their office and mobile phone, handheld, and notebook all synchronize their preset applications.

- *Internet access*: Access points and mobile phones can provide Internet access to handhelds and notebooks. These are essentially connection points to the wired LAN and not true wireless LANs.

- *Monetary and control transactions*: Pay parking fees, buy merchandise, and hotel check-in.

- *Cordless peripherals*: Examples are keyboards, mice, and headsets.

Bluetooth Low-Power Modes

Given that Bluetooth was developed for portable battery-driven products it is critical that the interface requires as little power as possible.

At the micro-level several provisions have been included to save power. First, the hopping mechanism is rather robust in that master and slave remain synchronized even if no packets are exchanged over the channel for several hundreds of milliseconds. No dummy data has to be exchanged to keep synchronization between master and slaves. Second, a receiver can decide quickly whether a packet is present or not. At the beginning of the receive slot, the receiver correlates the incoming signal in a sliding correlator which is matched to the access code. Since the access code only lasts for 70 μs, after a scan duration of about 100 μs, the receiver can decide to continue to listen or return to sleep. If the access code is not received within the scan window, no packet was sent or was so corrupted that further reception does not make sense.

The Bluetooth recipient can then sleep for the rest of the receive slot and transmit slot if the unit is a slave. If the proper access code is received the receiver will continue to demodulate the packet header. Checking the slave address in the header a slave can then determine whether the packet is intended for him. If not, he does not have to read the payload. Alternatively, if the slave address matches the packet type indication tells the slave whether there is a payload present and how long it may last.

At the macro-level Bluetooth offers different low-power modes for improving battery life. Piconets are formed when communication among devices is ready to take place. At all other times devices can be either turned off or programmed to wake up periodically to send or receive inquiry messages. When a Piconet is active slaves stay powered on to communicate with the master. It is possible to switch a slave into a low-power mode whereby it sleeps most of the time and wakes up only periodically. Three types of low-power modes have been defined:

1. *Hold mode* is used when a device should be put to sleep for a specified length of time. The master can put all its slaves in the hold mode to suspend activity in the current Piconet while it searches for new members and invites them to join.

2. *Sniff mode* is used to put a slave in a low-duty cycle mode, whereby it wakes up periodically to communicate with the master.

3. *Park mode* is similar to the sniff mode, but it is used to stay synchronized with the master without being an active member of the Piconet. The park mode enables the master to admit more than seven slaves in its Piconet.

1.3.4.2 Wireless LANs: An Overview

The popularity of Wi-Fi started back in 2000 after the ratification of the 802.11a, b standards. The acceptance of WLAN has spurred the growth of in-home networks and public hot spots as front ends to DSL or cable broadband pipes.

Today, many cities are in the process of deploying large-scale Wi-Fi networks, and in addition to those, there are hundreds of thousands of public hot spots that offer Wi-Fi service either for free or for a low charge. With so much Wi-Fi coverage available, it was only a matter of time until Wi-Fi made it into the cellular handset.

As Wi-Fi grew in the home, so did the number of people subscribing to broadband in the home. Many companies have taken advantage of this broadband popularity and started offering services where subscribers can carry on a phone voice call over a broadband connection. An extension of that same technology allows a voice call to be carried over a Wi-Fi connection.

There are two approaches within Spread Spectrum Techniques (SST): Frequency-Hopping Spread Spectrum (FHSS) and Direct Sequence Spread Spectrum (DSSS). Modulation is the process in which a lower-frequency wave is superimposed on a wave of higher-frequency that is fixed and constant (the carrier wave), thus modifying it to produce an information-bearing signal. DSSS "spreads" the signal across a broadband of radio frequencies and creates a redundant pattern called a "chip" to transmit. FHSS "hops" between frequencies as it encounters noise on the band. The type of modulation technique (DSSS or FHSS) used depends on the product. DSSS and FHSS are physical technologies and are not interoperable.

The delay spread of DSSS is caused by signals echoing off solid objects and arriving at the antenna at different times due to different path lengths. A band processor is needed to put the signal back together. The delay spread must be less than the symbol rate (the rate at which data is encoded for transmission). If the delay spread is greater than the symbol rate the signal spreads into the next symbol transmission ultimately slowing down the network. Coded Orthogonal Frequency Division Multiplexing (COFDM) transmits data in a parallel technique, as opposed to DSSS's "spread" and FHSS's "hopping" techniques, by slowing the symbol rate enough so that it is longer than the delay spread. The data is mapped across several lower-speed signals effectively blocking interference because transmission is spread across all sub-carriers in parallel. Table 1.3 shows a comparison of the three techniques.

Table 1.3: DSSS, FHSS, and CODFM Comparison

	DSSS	FHSS	COFDM
Band	2.4 GHz	2.4 GHz	5 GHz
Standard	IEEE 802.11b	IEEE 802.11b, Bluetooth	IEEE 802.11a
Carrier channel	Fixed in 17 MHz channel	Sends data over 1 MHz channel	20 MHz channel
Services supported	Data	Voice, data, video	Voice, data, video
Maximum number of independent channels	3	15	52 (48 data, 4 for error correction)

The frequency band is the speed at which the radio waves travel. Today there are two frequencies of choice for wireless LANs: 2.4 and 5 GHz. The 2.4 GHz band is used worldwide. However, in much of the world this band is used for cellular phones, creating interference.

The 2.4 GHz band has a power limit of 1 watt. Signals modulated at 2.4 GHz tend to reflect off of solid objects such as walls and buildings. Signal bouncing causes delays in delivery of data. While the 2.4 GHz band does not cause harmful interference, it does accept any interference it receives.

The 5 GHz band is also globally available. Operating in the 5 GHz band wireless LANs will not incur interference from Bluetooth devices, microwave ovens, cordless telephones, or wireless security cameras, since these devices cannot operate in this band. The IEEE 802.11a committee has ratified the standard for use of wireless devices in the 5 GHz band.

A comparison of alternative wireless LAN technologies is shown in Table 1.4.

1.3.4.3 IEEE 802.11 Standards

The IEEE 802.11 standard specifies the physical and MAC layers for wireless LANs. There are a number of 802.11 standards [7] including:

- *802.11a*: 5 GHz band via OFDM.

- *802.11b*: 2.4 GHz band, DSSS physical layer, and data transfer rates between 5.5 and 11 Mbps.

Table 1.4: Wireless Technology Alternatives

	IEEE 802.11b	Bluetooth
Frequency	2.4 GHz	2.4 GHz
Modulation technique	DSSS	FHSS
Maximum data rate	11 Mbps	1 Mbps
Voice support	No	Yes
Range	300′	30′
Security	40-bit encryption	128-bit authentication, 8 to 128-bit encryption
Transmit power	200 mW	20 mW

- *802.11d*: Defines physical requirements (channelization, hopping, patterns, new values for current Management Information Base (MIB) attributes) to extend into new regulatory domains (countries).

- *802.11e*: Enhancements to support LAN applications with Quality-of-Service (QoS), Class-of-Service (CoS), security, and authentication requirements such as voice, media streams, and video conferencing.

- *802.11f*: Recommended practices for Multi-Vendor Access Point Interoperability via Inter-Access Point Protocol (IAPP) Access Distribution Systems Support 802.11.

- *802.11g*: Use of DSSS to 20 Mbps and OFDM to 54 Mbps. This will be backward compatible with 802.11b and extend it to rates greater than 20 Mbps. This will improve access to fixed LANs and internetwork infrastructure (including other WLANs) via access points as well as ad hoc networks.

- *802.11n draft 2.0*: Rates of 250 Mbps at 200 m. Backward compatibility.

The 802.11b standard was developed to support roaming and wireless access in a large office and campus environment. The high data rates support high performance data networking applications such as wireless Internet and file sharing. The standard supports both frequency hopping and direct sequence. The hop rate of 802.11 is 2.5 Hz which is significantly less than that of Bluetooth and also makes it susceptible to interference. The standard is a true wireless LAN with significantly more power consumption and working range compared to Bluetooth.

1.3.4.4 Wi-Fi Power Pig

Wi-Fi devices have long been known to be power pigs, and much of this characteristic can be traced back to the early days of Wi-Fi, where performance and not battery life was the main goal. As Wi-Fi started entering more portable and smaller battery-powered devices, power has become a very important issue, but tackling the high power use of Wi-Fi technology has not been easy.

At first, a technology called Power Save Polling (PSP) was incorporated into Wi-Fi, but PSP only reduced the power that a Wi-Fi client required while on standby, and did nothing to reduce power while a handset is actually in use. PSP gave the access point the ability to power down a Wi-Fi client when it no longer needed to be communicated with. When the Wi-Fi access point needed to again communicate with a Wi-Fi client, it sent a signal that the client periodically checks for, and when the client received it, it woke up.

As more and more portable devices started using Wi-Fi, it became very apparent that something had to be done to increase the battery life of portable Wi-Fi devices, and solutions came out in both the underlying Wi-Fi standard, and in the chips that were used for Wi-Fi. To date, there never has been an extension to the 802.11 standard just for power saving measures (although this has been talked about), but there was an addition to the 802.11e standard that does attempt to reduce the power usage of portable Wi-Fi devices, especially voice over IP (VoIP) devices.

802.11e was primarily designed to add QoS to Wi-Fi, which is a way of prioritizing Wi-Fi packet traffic so that applications needing low latency (e.g. music, video, gaming, and voice traffic) will have a higher priority than bulk data traffic (e.g. file transfer) that is not time dependent. To address the low-power needs of portable Wi-Fi devices using QoS and 802.11e, an amendment was added to the standard that enhances the 802.11 legacy power save feature. This amendment, called Wi-Fi Multimedia (WMM) Power Save, is claimed to decrease power usage on a VoIP call from 15–40%. While that savings are not huge, it is a step in the right direction.

WMM Power Save works with cooperation from both the Wi-Fi access point and the Wi-Fi client. The Wi-Fi access point buffer packets that it normally would send to a Wi-Fi client, and instead of sending them to the client, it gives instructions to the client and tells it to go to sleep. After a short interval, the access point wakes up the client and sends it the data that it missed because it was sleeping. A Wi-Fi access point, a Wi-Fi client, and the software application running on the client device must all support WMM Power Save for the technology to be enabled and active.

The original 802.11 Power Save, along with the more recent WMM Power Save add small incremental improvements to the battery life of Wi-Fi portable devices, but for the larger improvements needed to make mobile Wi-Fi technology viable, much more is needed. This extra push has occurred by way of the chips that are used in Mobile Wi-Fi devices. Wi-Fi chipmakers know that for the technology to become popular, and thus for them to sell a lot of chips, the chips have to be very power thrifty.

1.4 Summary

Nomadicity refers to the system support needed to provide a rich set of computing, communications and multimedia capabilities and services, in a transparent and convenient form, to the nomad moving from place to place.

This new nomadic paradigm is already manifesting itself as users travel to many different locations with mobile devices like PMP, personal digital assistants, cellular telephones, pagers, and so on. This book addresses the open issue of power management support necessary for nomadicity.

Technological development and commercial innovation are breaking down traditional demarcations. The most often cited example is the camera phone; why have a phone and a camera when you can have both in one converged device?

In a converged world, content and access will no longer be in short supply. Future generations will be always connected to content, either via the Internet or other sources, and expect to access and consume content anywhere, anytime.

Broadband wireless and IP are examples of technology enablers that have defined a new generation of nomadic consumers.

Convergence is not restricted to physical devices, but also encompasses networks like IP Multimedia Subsystem (IMS), the associated person-to-person communications over the networks, and the content accessed over the networks.

For many years consumers have enjoyed devices for making phone calls, for digitally managing contacts and diaries, for listening to music, taking photographs, telling us where we are, for watching films and playing games.

The consumer now has a new dimension of choice, a choice between one-function devices, and these multi-functional devices. Unique combinations of capabilities that mix PMP with Wi-Fi for downloading content, PNDs with Bluetooth, Audio players with FM tuners just to name a few.

In addition to the convergence of mobile devices, applications, and services are also set to converge. Voice communications will be augmented with instant messaging, e-mail, video conferencing, VoIP, and mobile gaming. However, convergence is not a panacea. Combining a few capabilities on a mobile device may not lead to a commercial success. In many cases simple specialized models are preferred over multi-mode devices. The successful mobile devices will strike a balance between convergence and specialization.

As a result of these lifestyle changes battery-powered mobile devices are becoming more ubiquitous. However, not all growth is good. Concern over the environment is changing legislation and consumer behavior. Legislation will increase the cost of energy through laws against energy efficient obsolete technology, carbon pricing through taxation, and OECD countries recovering $150B per annum [8]. Below are some interesting facts on power [9]:

- World power consumption: 15 TWh

- US consumption/household: 11,000 KWh

- World average/household: 2,000 KWh

- Growth in consumption 2000–2004: 2 TWh

- Nuclear power plant output: 1 GW

- Hydroelectric power plants: 3–12 GW

- Portable devices daily usage: 0.5–1 GW

- Leakage of portable devices: 3B devices $\times$ 5 MW $=$ 15 MW

In addition to energy conservation, there are severe challenges with mobile devices. The cell phone has a limited talk time and may die in the middle of a conversation. The portable MP3 music player can store a collection of 500 h, but the batteries last less than 24 h. The mobile gaming device has a display that is very difficult to read. A superior gaming display would result in an unreasonably short battery lifetime. A laptop only works for a few hours; after that it just becomes an unusable brick. The GPS locator can only measure your position continuously for 4 h on one battery. The above examples illustrate that battery lifetime is an overall problem for portable devices. The following chapters address the creative solutions being applied to solve this energy problem.

References

[1] GSM/3G Market update, http://www.gsacom.com, June 2007.

[2] C. Ellis. The case for higher-level power management. In *Workshop on Hot Topics in Operating Systems*, pp. 162, 167, 1999.

[3] C. Chun and A. Barth. eXtreme energy conservation for mobile communications. *Freescale Technology Forum*, slide 3, July 2006.

[4] J. Ville and A. Vahervuori. *Energy-Awareness in Mobile and Ubiquitous Computing.* Department of Computer Science, University of Helsinki, Finland.

[5] B.A. Miller and C. Bisdikian. *Bluetooth Revealed*. Englewood Cliffs, NJ: Prentice Hall International, 2000).

[6] J.C. Haartsen. The Bluetooth radio system. *Personal Communications, IEEE (see also IEEE Wireless Communications)*, vol. 7, no. 1, pp. 28–36, 2000.

[7] WiFi Alliance, htpp://www.wifi.org, 2002

[8] Stern Review, Economics for Climate Change, 2007.

[9] International Energy Agency, http://www.iea.org, 2004.

CHAPTER 2

Hierarchical View of Energy Conservation

2.1 Issues and Challenges

High performance usually requires power sacrifices. The objective is to find the perfect balance between the two within a particular design. Optimize for performance when speed is an absolute must and target everything else for low power. There are a number of design and process strategies for achieving economical performance at system-level, chip-level, and even transistor-level designs, to achieve performance with long battery life.

2.1.1 Closing the Technology Gaps

Figure 2.1 summarizes the key challenges facing the mobile device industry. The step function labeled 1G, 2G, 3G, and 4G depicts the gains in cellular transmission over

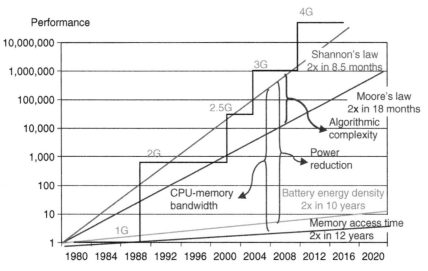

Figure 2.1: Key Technology Gaps [1]

time. This follows Shannon's law that predicts two times the transmission performance improvement in 8.5 months. Given Moore's law, it takes semiconductor manufacturers 18 months to double the number of transistors and therefore double the microprocessor performance. In addition, it takes battery makers 5–10 years to achieve comparable increase in power density. Also memory access time performance doubles every 12 years.

The gaps define the challenges faced by the mobile device industry. They include:

- Microprocessor and memory bandwidth gap
- Power reduction gap
- Algorithmic complexity gap

These gaps are the major hurdles to successful commercialization of mobile devices. In order to provide the advanced features and services required in future networks, system performance, as predicted by Shannon's law of Algorithm Complexity, must improve at a rate faster than Moore's law without compromising power budgets.

This has traditionally been tackled in the mobile device by making each instruction more efficient or executing multiple instructions at the same time. The increasing size of the Shannon–Moore gap with time means that incremental transistors and MHz alone are not sufficient to close this gap.

2.1.2 Always On, Always Connected: Paradox of the Portable Age

These "palm" size, portable devices that free people to go anywhere anytime also keeps them tethered by electrical power cords, plugs, and sockets. Sophisticated devices with multimode radios, color displays, 3D audio, video, and gaming features demand more from batteries without recharging. Users plunge from being always in touch to feeling impotent.

Users of portable devices have become "socket seekers." Strategically seeking out positions in airports, hotels, conference rooms, and home close enough to electrical sockets allowing one to recharge your portable devices.

The cycle of renewing battery life has introduced new rituals. Power strips are fully occupied with portable devices similar to animals feeding at a trough. Handheld and cell phones go into their cradles before you retire for the evening to bed. The digital cameras and iPod play musical chairs on the wall sockets. Commute time becomes critical charge time.

Well-dressed professionals can be found sitting next to whatever needs to be charged. Road warriors carry bags full of chargers. It is the most important and least talked about problem in consumer electronics.

Each year batteries become more powerful and circuitry improvements make devices more efficient. However, batteries cannot keep up with the rising expectations for longer life.

2.1.3 Balancing Battery Life with Performance and Cost

Mobile consumers have become accustomed to the size, weight, cost, and battery life of voice-only devices. New product offerings will be measured against these existing metrics, regardless of what new features they offer. Any noticeable regression from the current voice-only level could impact adoption on new data-centric devices. A mobile device that provides high speed data requires greater computing power and greater RF power consumption, resulting in shorter battery life.

In that respect, every technology has specific power requirements impacting battery life. Improvements in battery technology will enhance all radio access technologies, so that the differences between how current technologies use battery power are likely to persist for some time.

Figure 2.2 shows a comparison of the peak mobile power dissipation while transmitting for the different technologies. The values include both digital processing and RF elements.

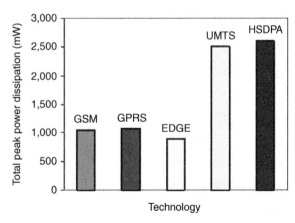

Figure 2.2: Power Consumption of Cellular Technologies

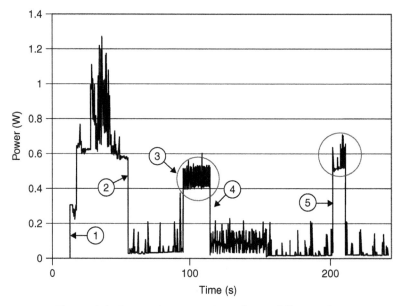

Figure 2.3: Power Consumption of a Mobile Handset

Source: www.Portelligent.com

Following the figure for specific cellular technologies, Figure 2.3 shows specifically when, during the operation of a mobile handset, power is consumed:

1. Turn power on – keypad and LCD backlights go on – search for network – welcome.

2. Keypad backlights go off – display goes dark.

3. Display goes on but keypad backlights stay off.

4. Display goes blank.

5. Close flip – external display backlight on.

2.2 Power versus Energy Types

Although the words power and energy are often used interchangeably in casual conversation, it is very important to understand the difference between these two concepts. Power and energy are defined in terms of the work the mobile device performs.

Power = Work/Time (Watts)
Energy = Power * Time (Joules)

The power used by a mobile device is the energy consumed per time unit. Conversely, energy is the time integral of power. Since a battery stores a given quantity of energy, the goal of energy management is to minimize the amount of energy required to perform each task satisfactorily.

In some cases, minimizing the power also minimizes energy. However, this is not always the case. Some tasks will require less energy to complete when being performed at high speed and high power for a short duration rather than at low speed and low power for a longer period of time.

Whether one should reduce the energy or power depends on the application. Fixed-duration tasks, such as playing audio or video, form an important application class where the energy required is directly proportional to the average power consumed as the duration of the task is constant. This class also includes waiting, whether waiting for user input when the device is on or keeping data in memory and the clock ticking when the device is off. For this class of tasks, focusing on minimizing power will minimize energy.

Therefore when talking about energy conservation, we need to distinguish between power reduction and energy reduction. There is a difference between energy and power.

Energy is an integration function of power over time. Just because we reduce power does not mean energy is reduced.

Here is a simple example to demonstrate this fact. When static power is negligible, for a given application that takes the same number of cycles to complete, whether the application runs fast and completes in a short amount of time, or runs at half frequency and takes twice as much time to complete, the same amount of energy is consumed.

However, when there is significant static power in the system, such as bias currents and leakage, running slow (or at a lower power) is less energy efficient due to this constant power being consumed during the entire active period.

The goal is to optimize systems for energy efficiency.

Power can be broken into two types: static and active power. Static power comes from any DC current sources such as reference circuits, analog designs, or even unwanted shorts in a product. Active power is from the switching activity of the circuitry where alpha is the switching activity.

Energy, on the other hand, is the integration of power over some period of time:

$$\text{Energy total} = \int \text{Power total } \mathrm{d}t$$

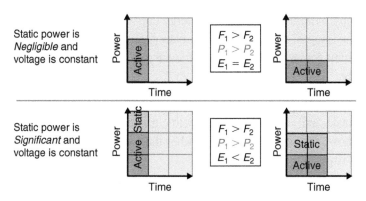

Figure 2.4: Energy versus Power [14]

- Energy
  - Integrated power over time
  - "Power reduction" ≠ "Energy reduction"

It is very likely that power reduction does not lead to energy reduction. One such case is if there is high static power and the frequency is reduced in half. The total power will be reduced, but since it takes so many clock edges to complete a function, if the frequency is reduced in half, it will take twice as long to complete which will increase the energy required to perform the task (Figure 2.4).

2.2.1 The Elements Power Consumption

This subsection will provide a general overview of the elements of power consumption.

Power consumption in complementary metal-oxide-semiconductor (CMOS) ICs can be abstracted to dynamic power (switching + short) plus static power. The primary elements of both dynamic and static power are shown in the following equation:

Power equation, P (avg) $= P$ (switching) $+ P$ (short) $+ P$ (static)

2.2.2 Elements of Dynamic and Static Power

Dynamic power consumption arises from circuit activity and is a result of switched capacitance and short-circuit current.

Dynamic or Active power $= P$ (capacitive switching) $+ P$ (short-circuit current)

P (switching) $=$ a Capacitive switching $=$ a $C(V)^{2}F$

- C = Capacitive load
- V = Supply voltage

- F = Switching frequency

- a = Switching activity

Switched capacitance is the primary source of dynamic power consumption and arises from charging and discharging of capacitors. Given the formula above there are 4 ways to reduce dynamic power consumption. These are discussed in subsequent chapters.

P (short-circuit) = $V * I$ short-circuit

P (short-circuit) = Crow-bar or transition current occurs when both the N channel and P channel are simultaneous active during switching creating a temporary short-circuit.

$$\text{Static power} = \text{DC power}$$

$$P \text{ (static)} = V * I \text{ leakage}$$

Static or leakage power consumption is the product of the supply voltage and the leakage current. Leakage power consumption is a result of leaking transistors and diodes and manifests itself in both active and standby modes of operation. In addition it is manufacturing process dependant. Static power includes sub-threshold leakage, reverse bias leakage and gate leakage currents.

2.3 Hierarchy of Energy Conservation Techniques

The approach to Energy conservation is a holistic approach which includes process technology, packaging, circuit and module design, System-on-a-Chip (SoC) design, tools, and system and application software to efficiently utilize the energy available to the system.

The last quarter century has seen enormous progress in the performance and capabilities of servers, desktop computers, laptops, and handheld devices. These gains have only increased the hunger for faster operation, greater functionality, lower prices, and smaller, more portable form factors.

Designing for energy-efficient performance is a paramount to ensure that new systems can support high-end applications without dramatically increasing energy consumption. This requires a fundamental rethinking on how to deliver new levels of performance within a given power envelope.

In order to deliver such energy-efficient platforms, a holistic effort is required across all common platform components including software, processors, hard drives, power supplies, radios, displays, and more. A holistic approach must include advanced power efficient micro architectures, industry-leading silicon technologies, and manufacturing expertise, world-class research, power-aware technologies, and unmatched ecosystem-building capabilities (packaging, software, tools, etc.). Table 2.1 indicates some the specific techniques applied at the various levels of design.

Table 2.1: Holistic Approach to Power Management

Holistic Approach Process Technology Transistor	Interconnects	Substrate	IC Circuit	Architecture	Packaging	System Software	Power Delivery and Management
Strained silicon	Low-k carbon doped oxide interlayer dielectric	Silicon on insulator	Body bias	Multicore and clustered architectures	Modules	Developer tools	Voltage regulation technology
Multi-Gate transistors	Copper		Dynamic sleep transistor	Multi-voltage islands	RCP	XEC	Improved display power space
High-k dielectrics			Demand base switching	Power gating		Dynamic voltage frequency scaling	Intelligent energy manager
Metal gates			Active power reduction	Clock gating			

A common approach to solve the complex power management problems that involve massive amounts of data is to break down the problem into manageable pieces and solve one piece at a time. If the pieces are small enough, the overall problem can be manageable. Of course, the trick is to bring all of the pieces together again and provide the answer to the original problem. This final integration stage is often overlooked and great implemented technology fails to deliver its promise due to an organizational oversight.

This approach has long been applied to complex engineering projects and is now finding its way into low power design. Hierarchical design can be roughly separated into three broad categories: planning, implementation, and assembly. Planning, often called "top-down" design, is the process of breaking the overall design into blocks that will be implemented individually. Planning is critical, because it sets the baseline for the entire project and must yield a final design that meets the project goals for timing, size, power, and other requirements (Figure 2.5).

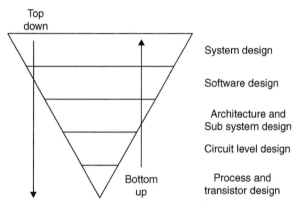

Figure 2.5: Hierarchical Design

Implementation and assembly, sometimes referred to as "bottom-up" design, is the process of implementing the detailed design of the individual blocks. Assembly is the process of connecting all of the blocks in the design to result in the final product. In this process, there are often preexisting IP blocks (such as memories, cores, etc.) that need to be considered during the planning process and added to the design during the assembly process. Also, it is more efficient to add "glue logic" during the assembly of the final chip, rather than structure it has a separate block that needs to be implemented during the bottom-up process.

Design teams develop and adopt hierarchical flows for many reasons. Complexity management and the need to shorten the turnaround time on large designs are usually at the top of the list. Using a hierarchical flow, the design can be attacked by different teams in different locations, resulting in concurrent engineering of the overall product. In addition, SoC design and intellectual property (IP) reuse methodologies also demand a hierarchical approach to chip design, because they involve bringing preexisting blocks into the design process. Figure 2.6 illustrates an example of an SoC block diagram and the various modules that comprise the SoC.

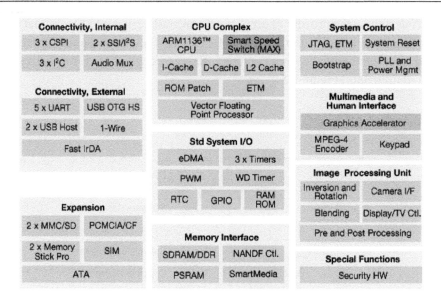

Figure 2.6: A Typical SoC Block Diagram

Source: i.MX Integrated Portable System Processor, www.freescale.com

For example, an SoC consists of the hardware, and the software that controls the, cores, peripherals, and interfaces. The design flow for an SoC aims to develop this hardware and software in parallel (see Figure 2.7).

Most SoCs are developed from pre-qualified hardware IP blocks together with the software device drivers that control their operation. The hardware blocks are put together using EDA tools and the software modules are integrated using a software integrated development environment.

A key step in the design flow is emulation: the hardware is mapped onto an emulator based on a field programmable gate array (FPGA) that mimics the behavior of the SoC, and the software modules are loaded into the memory of the emulation platform. Once programmed, the emulation platform enables the hardware and software of the SoC to be tested and debugged at close to its full operational speed.

After emulation the hardware of the SoC follows the place and route phase of the design of an integrated circuit before it is fabricated.

SoCs are verified for functional correctness before being sent to the wafer fab. This process is called verification. Hardware Descriptive Languages are used for verification. With growing complexity of SoCs, Hardware Verification Languages like System

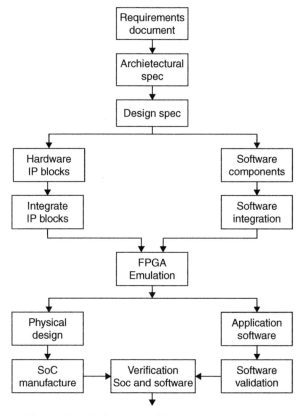

Figure 2.7: Software and Hardware Co-Design

Verilog and System C, are used. The bugs found in the verification stage are reported to the designer. Traditionally, 70% of time and energy in SoC design life cycle are spent on verification. A deeper look at certain aspects of low power SoC deign, tools, and standards is discussed later in this book.

There are other less compelling reasons for adopting hierarchical design that are a result of the limitations of today's conventional point-tool flows. Many design teams have been forced to turn to hierarchy simply because of the limitations in their tool flow. For example, the market-leading logic synthesis system is limited to handle blocks of gates. In other cases, hierarchical design is applied as brute force to manage the critical timing in a design by ensuring that certain design blocks are placed together. This is not an optimal approach, because it sacrifices the big picture view of the design while attempting to control only small portions of the design.

The following sections address the hierarchical approach to low power design of portable devices starting with process and transistor technology.

2.4 Low Power Process and Transistor Technology

The genesis of low power electronics can be traced to the invention of the bipolar transistor in 1947. Elimination of the suppressive requirements for several watts of filament power and several hundred volts of anode voltage in vacuum tubes in exchange for transistor operation in the tens of milliwatts range was a breakthrough of unmatched importance in low power electronics. The capability to fully exploit the superb low power assets of the bipolar transistor was provided by a second breakthrough, the invention of the integrated circuit in 1958 (Figure 2.8).

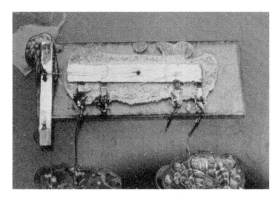

Figure 2.8: First IC Made by Jack Kilby of Texas Instruments in 1958

Source: www.ti.com

Although far less widely acclaimed as such, a third breakthrough of indispensable importance to modern low power digital electronics was the CMOS integrated circuit announced in 1963. Collectively, these three inventions demonstrated the critical conceptual basis for modern low power electronics.

2.4.1 Process Technology Scaling

The semiconductor industry has been built on the benefits of traditional CMOS scaling. Classical scaling shrinks voltages along with the lithographic dimensions, and the results were that for a given die size, scaling delivered more gates, switching faster, at a constant power.

Scaling advanced CMOS technology to the next generation improves performance, increases transistor density, and reduces power consumption. Technology scaling typically has three main goals:

1. Reduce gate delay by 30%, resulting in an increase in operating frequency of about 43%.

2. Double transistor density.

3. Reduce energy per transition by about 65%, saving 50% of power (at a 43% increase in frequency).

Unfortunately, some parameters have hard limits in scaling. A key component of the performance of CMOS transistors is the drive current or "I_{on}." This performance is proportional to the gate overdrive, or basically the supply voltage minus threshold voltage (V_t). If we continue scaling the supply voltage then the threshold voltage must also scale to keep up with the gate overdrives or to keep the gate overdrive up. CMOS transistors are not perfect switches. They have an off-state current that rises exponentially as the threshold voltage is reduced and so while lithographic dimensions continue to scale, voltage scaling itself has slowed down. The new realities of scaling are impacting both the static and dynamic power associated with products.

2.4.1.1 Scaling Theory

Scaling a technology reduces gate delay by 30% and the lateral and vertical dimensions by 30%. Therefore, the area and fringing capacitance, and consequently the total capacitance, decrease by 30% to 0.7.

2.4.1.2 Power Consumption

A chip's maximum power consumption depends on its technology as well as its implementation. According to scaling theory, a design ported to the next-generation technology should operate at 43% higher frequency.

If the supply voltage remains constant (constant voltage scaling), the power should remain the same. On the other hand, if the supply voltage scales down by 30% (constant electric field scaling), the power should decrease by 50%:

$$V_{DD} = 1 \text{ (constant voltage scaling)}$$

$$\text{Power} = C * V^2 * f = 0.7 * 1 * (1/0.7) = 1$$

$$V_{DD} = 0.7 \text{ (constant electric field scaling)}$$

$$\text{Power} = C*V^2*f = 0.7*0.72*(1/0.7) = 0.5$$

Unfortunately, logic designs ported to next-generation technologies with constant voltage scaling do not show a decrease in power; the power remains constant. On the other hand, logic designs ported to technologies using constant electric field scaling decrease in power. This is consistent with scaling theory.

However, the power dissipation of a chip depends not only on its technology, but also on its implementation, i.e. on size, circuit style, architecture, operation frequency, and so on.

Constant electric field scaling (supply voltage scaling) gives the lower energy delay product (ignoring leakage energy) and hence is preferable. However, it requires scaling threshold voltage (V_t) as well, which increases the sub-threshold leakage current, thus increasing the chip's leakage power.

2.4.1.3 Scaling Paradox

A consequence of scaling is rather undesirable, but until recently it has not been a particularly negative feature; the standby current density increases exponentially as the length scale is decreased. Figure 2.9 illustrates the passive power trend based on sub-threshold currents

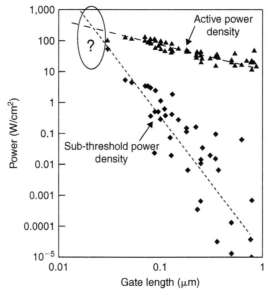

Figure 2.9: Active (Dynamic) versus Static Leakage [2]

calculated from the industry trends of V_t, all for a junction temperature Tj = 25°C. For reference, the active power density is copied onto this scale to illustrate that the sub-threshold component of power dissipation is emerging to compete with the long-battled active power component for even the most power tolerant, high-speed CMOS applications.

As the lithography pushes forward, the product designer must devise new strategies to cope with the interference of passive power, which pushes for higher V_t (and thus higher V_{DD}) versus active power, which demands lower V_{DD} and thus lower V_t. This results in fragmentation of device design points that address these conflicting needs in wafer fabrication.

The approach outlined in Table 2.2 allows the product designer flexibility to choose the best device match for active and passive power versus performance. Products that are very sensitive to passive power, such as portable and handheld devices, may sacrifice some performance to enable higher V_t. If these designs require higher performance, they are forced to sacrifice some switching power by use of correspondingly higher V_{DD} as well.

Table 2.2: Wafer Suppliers Offer a Variety of Process Nodes to Meet Different Applications [2]

Application	High Performance	1.2-V logic	1.5-V logic	Low Power	Interface
V_{DD} (V)	1.2	1.2	1.5	1.2	2.5
T_{OX} (nm)	1.8	2.2	2.2	2.2	5
$I_{D\text{-}OFF}$ (nA/μm)	10	3	6	0.05	0.01

Other applications may be challenged to inexpensively conduct heat generated by active power away from the integrated circuits and thus favor lower-V_{DD} devices with low V_t and higher passive power. Thus, the variety of threshold voltages and power-supply voltages offered in 130 nm technology has expanded to address these diverse needs.

Managing power as technology is scaled is a continual interaction between the product design needs and the physical constraints of the process technology. Figure 2.10 gives a good indication of those relationships.

Given a speed constraint, the goal is to find is the optimal supply voltage and the transistor threshold voltage while minimizing the power. Since these choices impact tradeoffs between static and dynamic power, then the application-dependent power constraints need to enter into that optimization level. Once a performance target has been set, a simple back-of-the-envelope analysis shows that the optimal supply voltage and the optimal threshold voltage choices imply a static power that is about 30% of the total power. The best choice depends on the average active power, which is strongly dependent on the circuit activity.

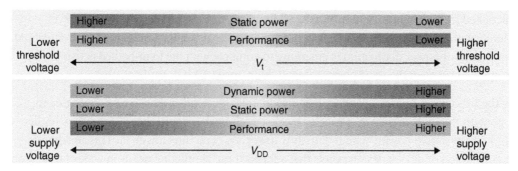

Figure 2.10: Threshold Voltage Tradeoff

Source: www.freescale.com

2.4.2 Transistors and Interconnects

2.4.2.1 Planar Transistor

Planar transistors have been the core of integrated circuits for several decades, during which the size of the individual transistors has steadily decreased. Rapid and predictable miniaturization was predicted by Moore's law [3]; this has allowed the semiconductor industry to make new products with added functions with each new generation of technology (Figure 2.11).

This predictable scaling is now reaching its limit [4] and has forced the industry to look to novel transistor architectures beyond the 45 nm technology node. At such sizes, planar transistors suffer from the undesirable short channel effects, in particular "off-state" leakage current which increases the standby power required by the mobile device.

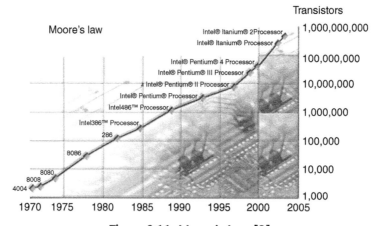

Figure 2.11: Moore's Law [3]

In all these years of digital CMOS innovation and scaling, we have only scratched the surface of the semiconductor substrate. The Planar CMOS device, the workhorse of digital applications used in modern electronic systems, has a channel only on the surface of the silicon. These devices have a single gate on the surface of the silicon to modulate the channel on the surface of the semiconductor. Scaling of these planar devices has now begun to hit its limits for power, noise, reliability, parasitic capacitances, and resistance. New device architectures using multiple sides of the semiconductor, not just the planar surface, offer a path to overcome these performance limits. In addition, these non-planar CMOS devices enable new circuits previously not possible with single gate CMOS devices [5–7].

2.4.2.2 CMOS Switch

The fundamental function of a transistor in a digital system is a switch: to conduct as much current as possible when on, and to shut down when off. The limits of planar CMOS technologies make this fundamental operation impractical as gate lengths and supply voltages are scaled down. The current in the on state is reduced when device sizes are scaled down, due to reduced mobility of the electrons and parasitic resistances (among other effects). The leakage current increases when the device is turned off; this substantially higher leakage can drain batteries quickly, making many mobile applications difficult to engineer (Figure 2.12).

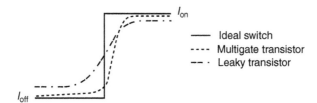

Figure 2.12: Ideal Switch Compared to a Good Switch Compared to a Leaky Switch

The fundamental limiting factors to scaling a single gate planar CMOS transistor are the leakage through the gate, and the effect of the drain taking control of the channel making it difficult to control the switch using the gate (Figure 2.13); this is known as short

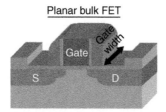

Figure 2.13: Planar MOSFET

Source: www.freescale.com

channel effects. In a multi-gate device, the channel of the device is controlled (gated) by gates from multiple sides and the body of the device where the channel is formed is made ultra-thin, so that the gate bias controls the channels more efficiently from multiple sides (Figure 2.14).

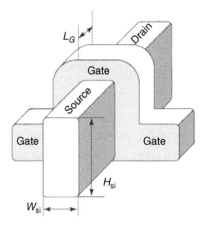

Figure 2.14: Tri-Gate Transistor, Gates Surround the Silicon Channel on Three of Four Sides

Source: www.intel.com

These multi-gate devices have a single gate electrode, but this single gate electrode wraps around many sides and controls the channel from multiple sides. These have various names, such as, FinFET, Tri-GATE, IT Gate, etc. In all these devices, a single gate electrode controls the channel from multiple sides yielding better control of the device and lower leakage when it is shutdown and conducts more current when turned on.

2.4.2.3 Multi-Gate Transistors

In a multi-gate device, the channel is surrounded by several gates on multiple surfaces, allowing more effective suppression of "off-state" leakage current. Multiple gates also allow enhanced current in the "on" state, also known as drive current. These advantages translate to lower power consumption and enhanced device performance. Non-planar devices are also more compact than conventional planar transistors, enabling higher transistor density which translates to smaller overall microelectronics.

Dozens of multi-gate transistor variants may be found in the literature. In general, these variants may be differentiated and classified in terms of architecture (planar versus non-planar design) and number of channels/gates (2, 3, or 4).

2.4.2.4 Planar Double-Gate Transistors

Planar double-gate transistors employ conventional planar (layer-by-layer) manufacturing processes to create double-gate devices, avoiding more stringent lithography requirements associated with non-planar, vertical transistor structures. In planar double-gate transistors the channel is sandwiched between two independently fabricated gate/gate oxide stacks. The primary challenge in fabricating such structures is achieving satisfactory self-alignment between the upper and lower gates.

2.4.2.5 FinFETs

The term FinFET was coined by University of California, Berkeley researchers to describe a non-planar, double-gate transistor built on an SOI substrate, based on the earlier DELTA (single-gate) transistor design. The distinguishing characteristic of the FinFET is that the conducting channel is wrapped around a thin silicon "fin," which forms the body of the device. The dimensions of the fin determine the effective channel length of the device.

In the technical literature, FinFET is used somewhat generically to describe any fin-based, multi-gate transistor architecture regardless of number of gates.

A 25-nm transistor operating on just 0.7 V was demonstrated in December 2002 by Taiwan Semiconductor Manufacturing Company. The "Omega FinFET" design, named after the similarity between the Greek letter "Omega" and the shape in which the gate wraps around the source/drain structure, has a gate delay of just 0.39 picosecond (ps) for the N-type transistor and 0.88 ps for the P-type transistor (Figure 2.15).

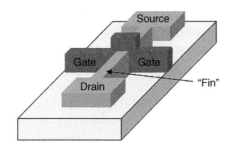

Figure 2.15: A Double-Gate FinFET Device

Source: www.answers.com

2.4.2.6 Tri-Gate Transistors

Tri-gate or 3-D is the term used to describe non-planar transistor architecture planned for use in future microprocessor technologies. These transistors employ a single gate stacked on top of two vertical gates allowing for essentially three times the surface area for electrons to travel. It has been reported that their tri-gate transistors reduce leakage and consume far less power than current transistors.

In the technical literature, the term tri-gate is sometimes used generically to denote any multi-gate FET with three effective gates or channels (Figure 2.16).

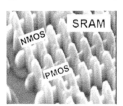

Figure 2.16: SEM View of SRAM Tri-Gate Transistors

Source: www.intel.com

2.4.2.7 Inverted T Gate Transistors

Multi-gate device architectures are rapidly evolving. The FinFET, with all its advantages, still has a significant drawback: the region between the fins is not used as part of the switch. A new family of devices called ITFET that uses both the vertical and horizontal regions of silicon has been proposed and demonstrated. The ITFET has both vertical and horizontal thin body regions shaped like an "inverted T." The ITFET offers maximum surface area utilization on the wafer for the channel and allows optimization of crucial circuit elements such as the SRAM-based cache that is ubiquitous in all modern digital CMOS products [8, 9] (Figure 2.17).

Bottom pedestal offers mechanical support for thin and high aspect ratio FIN providing better scaling and yield improvement. Also allows for device width "tuning" instead of FinFET's discrete width increment which is particularly good for SRAM design. In addition, better source/drain contact that results in reduced parasitic resistance.

2.4.2.8 Process Techniques

The multi-gate device architecture requires two basic technologies that are substantially new compared to existing processes:

1. A process technology to make very thin silicon body of the order of 20 nm.
2. A process to fabricate identical gates on at least two sides of this very thin silicon.

Various process technologies have been proposed to fabricate such a structure (Figure 2.18). While many processes have been identified to make a very thin silicon channel, a process that easily allows gates on both sides of this channel that are aligned to each other has been provided only on the structure now called FinFET, Tri-Gate etc. Currently, most research

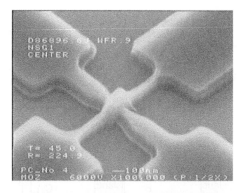

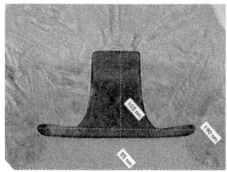

Figure 2.17: The ITFET Has a Channel that is Shaped Like an Inverted T

Source: www.freescale.com

efforts to make multi-gate devices involve these FinFET device structures. While devices with sub-20 nm silicon body and gates less than 40 nm have been demonstrated, there are still manufacturing challenges to make a product with millions of such transistors [10–14].

While these novel devices make progress, new gate materials are also researched. Incorporating these new materials is crucial to gain the maximum benefits out of these novel structures. The use of metal gates instead of conventional polysilicon gates will allow less parasitic resistance and poly depletion effects. Patterning these metal gates on FinFETs with traditional oxides is a challenge, but they have been demonstrated with new process techniques (Figure 2.19).

2.4.2.9 Transistor Design

Figure 2.20 highlights the top five areas of focus on transistor design to improve the power/performance tradeoff. These areas are key for the transistor development delivering on performance improvements while minimizing the leakage currents and parasitics that reduce the performance or increase the power.

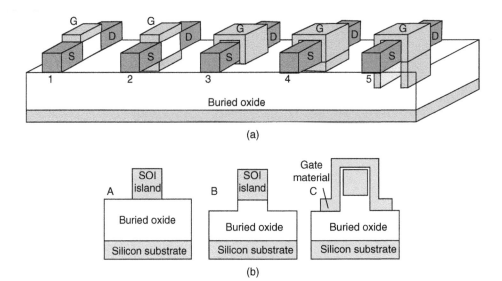

(a)

(b)

Figure 2.18: Multiple-Gate MOSFETs [15]. (a) Different Gate Configurations for SOI Devices: 1, Single Gate; 2, Double Gate; 3, Triple Gate; 4, Quadruple Gate; 5, New Proposed Pi-Gate MOSFET. (b) Basic Pi-Gate Fabrication Steps (Cross-Section): A, Silicon Island Patterning; B, Shallow Buried Oxide RIE; C, Gate Oxide Growth and Gate Material Deposition and Patterning

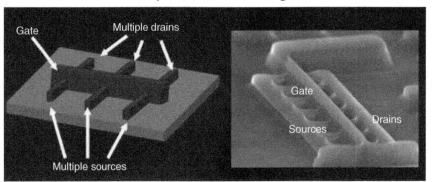

Figure 2.19: Schematic and SEM View of a Tri-Gate Transistor

Source: www.intel.com

2.4.2.10 Strained Silicon

When silicon is allowed to deposit on top of a substrate with atoms spaced farther apart than in silicon lattice, the atoms in silicon are naturally stretched to align with the substrate atoms underneath, leading to "strained silicon." Moving these silicon atoms

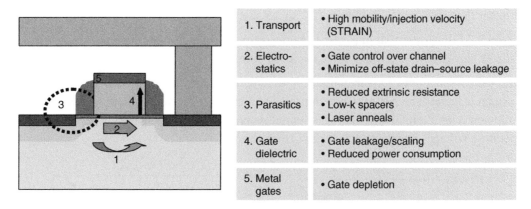

1. Transport	• High mobility/injection velocity (STRAIN)
2. Electro-statics	• Gate control over channel • Minimize off-state drain–source leakage
3. Parasitics	• Reduced extrinsic resistance • Low-k spacers • Laser anneals
4. Gate dielectric	• Gate leakage/scaling • Reduced power consumption
5. Metal gates	• Gate depletion

Figure 2.20: Top 5 Areas for Transistor Development

Source: www.freescale.com

farther apart reduces the atomic forces that interfere with the movement of electrons through the transistors, subsequently resulting in better chip performance and lower energy consumption, e.g. the electrons can flow 70% faster in strained silicon, leading to chips that can be 35% faster in performance. Strained silicon strategy involves either inserting germanium atoms into silicon lattice or adding intermediate layer of silicon-germanium compound semiconductor.

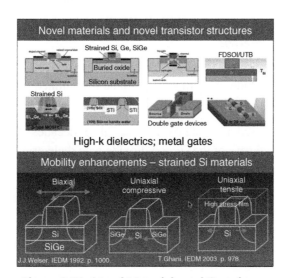

Figure 2.21: Novel Materials and Transistors

Source: T. Ghani, IEDM 2003, p. 978

Strained silicon engineering is a technique used to increase the transistor channel mobility of either the nFET or pFET. Figure 2.21 shows a couple of examples of strain engineering.

Figure 2.22 illustrates that improving the device performance starts with managing the tradeoff between drive current versus leakage current.

Strained silicon allows for substantial improvements in the transistors performance at a constant leakage. The performance improvement can be achieved without increasing the static power or alternatively, static and dynamic power can be decreased by lowering the operating voltage, given a constant performance goal.

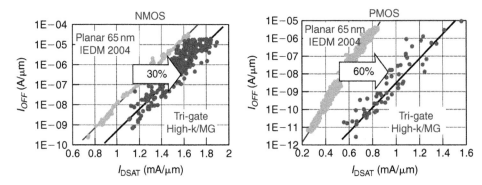

Figure 2.22: Integrated Tri-Gate NMOS and PMOS Transistors Demonstrate Record Drive Current Performance

Source: www.intel.com

2.4.2.11 High-k Gate Dielectric

The term high-k dielectric refers to a material with a high dielectric constant (k), as compared to silicon dioxide, used in semiconductor manufacturing processes which replaces the silicon dioxide gate dielectric.

Silicon dioxide has been used as a gate oxide material for decades. As transistors have decreased in size, the thickness of the silicon dioxide gate dielectric has steadily decreased to increase the gate capacitance and thereby drive current and device performance.

The problem is that as the oxide gets thinner, the rate of gate leakage tunneling goes up contributing to power dissipation and heat.

As the thickness scales below 2 nm, leakage currents due to tunneling increase drastically, leading to unwieldy power consumption and reduced device reliability. Replacing the silicon dioxide gate dielectric with a high-k material allows increased gate capacitance without the contaminant leakage effects. In addition to its dielectric properties silicon dioxide has an almost defect free dielectric interface which insures good compatibility with the silicon substrate and also the polysilicon gate electrode.

Conventional silicon dioxide gate dielectric structure compared to a potential high-k dielectric structure (Figure 2.23).

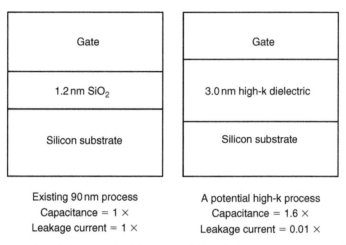

Existing 90 nm process
Capacitance = 1 ×
Leakage current = 1 ×

A potential high-k process
Capacitance = 1.6 ×
Leakage current = 0.01 ×

Figure 2.23: Conventional Silicon Dioxide Gate Dielectric Structure Compared to a Potential High-k Dielectric Structure

The gate oxide in a MOSFET can be modeled as a parallel plate capacitor. Ignoring quantum mechanical and depletion effects from the silicon substrate and gate, the capacitance C of this parallel plate capacitor is given by:

$$C = \frac{\kappa \varepsilon_0 A}{t}$$

where

- A is the capacitor area

- κ is the relative dielectric constant of the material (3.9 for silicon dioxide)

- ε_0 is the permittivity of free space

- t is the thickness of the capacitor oxide insulator

Since leakage limitation constrains further reduction of *t*, an alternative method to increase gate capacitance is alter k by replacing silicon dioxide with a high-k material. In such a scenario, a thicker gate layer might be used which can reduce the leakage current flowing through the structure as well as improving the gate dielectric reliability (Figure 2.24).

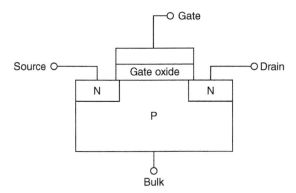

Figure 2.24: Cross-Section of an Nch MOSFET Showing the Gate Oxide Dielectric

2.4.2.12 Gate Capacitance Impact on Drive Current

The drive current I_D for a MOSFET can be written, using the gradual channel approximation, as:

$$I_D = \frac{W}{L} \mu C_{inv} \left(V_G - V_t - \frac{V_D}{2} \right) V_D$$

where

- *W* is the width of the transistor channel
- *L* is the channel length
- μ is the channel carrier mobility (assumed constant here)
- C_{inv} is the capacitance density associated with the gate dielectric when the underlying channel is in the inverted state
- V_G is the voltage applied to the transistor gate
- V_D is the voltage applied to the transistor drain
- V_t is the threshold voltage

It can be seen that in this approximation the drain current is proportional to the average charge across the channel with a potential $V_D/2$ and the average electric field V_D/L along the channel direction. Initially, I_D increases linearly with V_D and then eventually saturates to a maximum when:

$$V_{D,\text{sat}} = V_G - V_t$$

to yield:

$$I_{D,\text{sat}} = \frac{W}{L} \mu C_{\text{inv}} \frac{(V_G - V_t)^2}{2}$$

The term $(V_G - V_t)$ is limited in range due to reliability and room temperature operation constraints, since too large a V_G would create an undesirable, high electric field across the oxide. Furthermore, V_t cannot easily be reduced below about 200 mV, because κT is approximately 25 mV at room temperature. Typical specification temperatures $<100°C$ could therefore cause statistical fluctuations in thermal energy, which would adversely affect the desired V_t value. Thus, even in this simplified approximation, a reduction in the channel length or an increase in the gate dielectric capacitance will result in an increased $I_{D,\text{sat}}$.

Replacing the silicon dioxide gate dielectric with another material adds complexity to the manufacturing process. Silicon dioxide can be formed by oxidizing the underlying silicon, ensuring a uniform, conformal oxide, and high interface quality. As a consequence, development efforts have focused on finding a material with a requisitely high dielectric constant that can be easily integrated into a manufacturing process. Other key considerations include band alignment to silicon (which may alter leakage current), film morphology, thermal stability, maintenance of a high mobility of charge carriers in the channel, and minimization of electrical defects in the film/interface. Materials which have received considerable attention are hafnium and zirconium silicates and oxides, typically deposited using atomic layer deposition.

Rapid and predictable scaling of planar CMOS devices is becoming difficult. New device structures to replace planar CMOS devices are being researched. Multi-gate devices using multiple surfaces are promising continued scaling, and could even make new circuits feasible. These devices can provide new and better characteristics across all logic, memory, and analog device function. The challenges in making these devices to enter mainstream products are many. However, rapid strides in process, design, and modeling in the last few years has delivered substantial progress.

2.4.2.13 New Logic Circuits with a Better Switch

Traditional digital CMOS has operated between ON and OFF states, as explained in Figure 2.12. The excellent I_{on} and I_{off} characteristics of the multi-gate devices allow future scaling of traditional circuits for a few generations. Even this is not sufficient for certain low-power applications such as pacemakers, hearing aids, and other self-powered logic devices. While sub-threshold logic has been proposed as a low-power circuit alternative, it has not been widely used, due in part to the limitations of single gate devices. Multi-gate devices, with their steep turn on characteristics and extremely low leakage characteristics, promise to be ideal to make these systems practical [16].

2.4.2.14 New Analog Circuits Can Use Multiple Independent Gates

While digital CMOS logic leads the process technology roadmap for computing applications, the communications or RF applications have a substantial mix of analog components that are integrated into the CMOS logic or as standalone products. The double-gate device architecture allows better scaling of these analog applications and new functions that were not possible with single gate transistors. Just as in CMOS logic, the fundamental switch is improved by the double-gate architecture for analog applications. The double-gate architecture offers better gain and can be used as a better mixer, amplifier or VCO.

2.4.2.15 Beyond the Transistor

Rotated substrates or junction engineering are examples of developments to deliver the best drive currents while minimizing the off-state leakages. SOI technology brings higher speed at lower power through reduced capacitances and improved performance at a given voltage. High-k Metal-Insulator-Metal (MIM) capacitor is a decoupling capacitance technology that helps voltage supply integrity. And also, the low-k and ultra-low-k interconnect insulators reduce capacitance and thereby drive down the chip-level power.

2.4.2.16 Copper Interconnects and Low-k Dielectrics

With continuing device scaling, wiring interconnect becomes increasingly important in limiting chip density and performance. Fundamental changes in interconnect materials are needed with Copper (Cu) replacing aluminum and low permittivity dielectrics replacing silicon dioxide. The integration of these two advanced materials results in significant reduction in signal delay, cross-talk, and power dissipation, enabling the semiconductor industry to continue device scaling.

The fabrication of Cu low-k interconnects requires novel materials and processes, including electroplating Cu, dual damascene structures, chemical–mechanical polishing, ultra-thin barriers, and passivation layers. These novel materials and processes give rise to distinct structure and defect characteristics raising yield and reliability concerns for Cu/low-k interconnects.

As the technology continues to advance, the implementation of porous low-k dielectrics beyond the 65 nm node will bring in new processing and reliability issues, such as pore sealing, etch damage, and ultra-thin barriers. These problems will be discussed together with recent advances in material and process development and reliability improvement for Cu low-k interconnects.

2.4.2.17 Dual Voltage Threshold Transistor

Each transistor in a semiconductor design has an associated threshold voltage (V_t) that determines at what voltage level the transistor is triggered to open or close. Typically, a lower V_t transistor offers higher performance because the voltage does not have to swing as far to trigger the transistor (Figure 2.25).

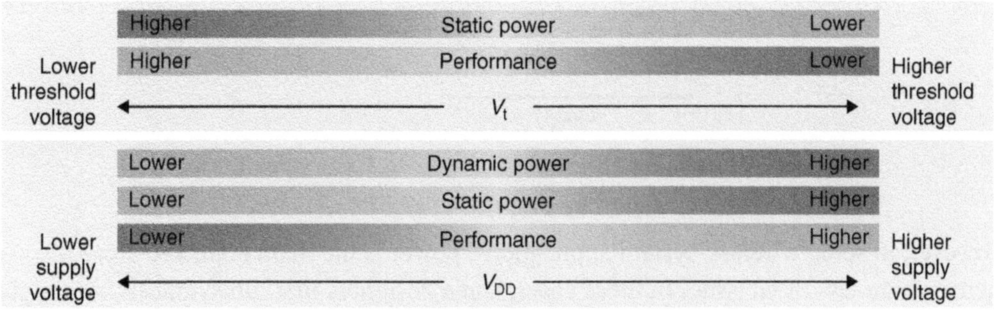

Figure 2.25: Threshold Voltage Tradeoff

Source: www.freescale.com

Design libraries include low-threshold-voltage, high-threshold-voltage, and threshold-voltage-with-MTCMOS (multi-threshold CMOS) cells. Multiple-cell libraries help designers to deal with both leakage and dynamic power. One library that is going to work for an entire design, because you have designs that are speed-critical, and, for the areas that are not speed-critical, you want to reduce the leakage.

A multi-cell library typically comprises at least two sets of identical cells that have different threshold voltages. Those with higher threshold voltage are slower but have less leakage; conversely, the cells with lower threshold voltage are faster but leak. Conceding a little bit of speed, you get a reduction in leakage. A high-threshold-voltage cell typically has 50% less leakage than a low-threshold-voltage cell with no bad side effects, such as area gain.

For most applications, designers typically use a low-threshold-voltage library for a first pass through synthesis to get maximum performance and meet timing goals. They then determine the critical paths in their design, that is, the path or paths in the design that require the highest performance. They then try to locate areas that do not require low-threshold-voltage cells and swap out low-voltage cells for high-voltage cells to reduce overall power and leakage of the design. This approach represents the most common use of the multi-threshold design technique because most applications have timing as a first requirement, low-threshold-voltage libraries run faster through synthesis, and synthesis tools ultimately produce smaller design areas from these libraries. Synthesis tools tend to run longer and produce larger design areas when running heavy doses of high-threshold-voltage cells (Figure 2.26).

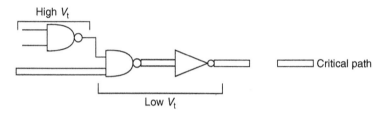

Figure 2.26: Mixing Threshold Voltages to Ensure Performance and Power Targets Are Met

However, in some wireless-system applications, power is the main goal, and area increases are less of an issue. In those cases, some designers first run synthesis with high-threshold-voltage cells, find the critical path, and then swap out the high-voltage cells with low-voltage cells until they reach their performance goal.

2.4.2.18 Active Well Biasing

Essentially, standby current drain is electrons "leaking" through a gate junction of thin oxide within the transistor. The amount of current flow is a function of the voltage across the junction. The greater the difference between voltages on both sides of the junction, the greater the leakage, more current is being drained, even in standby mode.

Well-biasing techniques help control channel leakage currents. Active well bias enhances the energy/performance relationship by manipulating transistor performance real-time. With an active back bias technique low V_t device is used for maximum performance, and then raise the threshold in standby mode, thus providing two performance levels in one transistor.

Active well bias affects the threshold voltage, which in turn affects the sub-threshold leakage currents. If the body of a PMOS device is increased above V_{DD} or the body of an NMOS device is reduced below ground, the device is said to be in back bias and the sub-threshold leakage current can be reduced.

2.5 Low Power Packaging Techniques

2.5.1 Introduction

As mobile devices perform more in a smaller form factor, the technologies that enable them are challenged to do the same. As SoC vendors integrate more functionality into slimmer, smaller form factors, advanced packaging, along with SoC integration, is used to reduce the size, cost, and power consumption of the SoC's inside the mobile devices. New technologies, such as multi-die System-in-Package (SiP), stacked packages or Package-on-Package (PoP), packages integrating other packages or Package-in-Package (PiP), packages that integrate passives with silicon die, all promise to improve the power consumption, performance, and reduce package sizes. These innovations in SoC packaging technology enable vendors of converged products, such as cellular smartphones and portable media players, to create the small, thin, multi-function devices to meet their market demands.

The challenges of combining digital, analog and RF functions into a single piece of silicon and optimizing this for different process technologies has proved difficult and extremely costly. Although major advances are continuing to be made in the semiconductor industry, for highly complex systems containing multi-functional components (i.e. digital, analog, RF, MEMS, optics, etc.), the SoC option will be very costly if it can be achieved at all.

Dramatic increases in wireless communication and application processing, the number of radio interfaces, and the amount of memory integrated into advanced handsets are overwhelming compared to simple, voice-only designs. At the same time, handset users expect very small, sleek, low-cost handsets that feature large color displays and standby and talk times similar to those of voice-only handsets. In response, handset component manufacturers are moving aggressively to include advanced power-management techniques, specially tailoring communication and application processing architectures to the required tasks and dramatically reducing component cost.

The three main approaches to reduce component cost are SoC integration, SiP integration (the stacked-die solution), and PoP integration (the stacked package solution). System integration requires a complete system-level tradeoff analysis that involves the entire bill of materials and the architectures used for the communication system being designed. Determining the best set of trade-offs to attain the most cost-effective solution requires close collaboration from diverse engineering disciplines, including process integration engineers, circuit engineers, and system engineers.

Figure 2.27 illustrates how SoC [17] for highly complex systems can become prohibitively expensive. It also details the different waves of packaging that have helped to produce decreases in system size and how this has coincided with Moore's law. Innovative packaging technologies like SiP are viewed as a critical enabler in helping continue this trend and fill what is called the **packaging gap**.

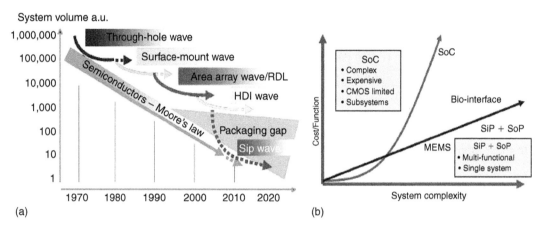

Figure 2.27: (a) Packaging Waves (1970–2020) and (b) Costs of SiP versus SoC

Source: IZM Fraunhofer

2.5.2 Systems-in-Package

SiP technology provides a complete functional system in one module that can be processed much like a standard component during board assembly. In contrast with SoCs, where a single die is used, SiP modules are usually integrated by stacking dies vertically one upon the other, or combining multiple die stacks onto the same substrate, or even embedding die structures into the substrate. This is in contrast to the horizontal nature of SoC, which overcomes some of the SoC limitations, such as latency if the size of the chips and their thicknesses used in stacking are small. Figure 2.28 illustrates these different SiP technologies.

2.5.3 Package-on-Package

Package-on-Package takes this integration a step further than SiP, placing one package on top of another for greater integration complexity and interconnect density. PoP also enables procurement flexibility, lower cost of ownership, better total system and solution costs and faster time to market.

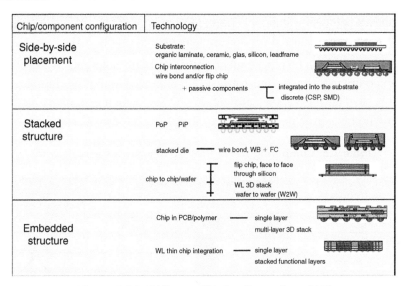

Figure 2.28: Different SiP Configurations [13]

Figure 2.29 illustrates that the top component in this kind of package can be any SoC, such as memory or a graphics processor which is added to the main processor to produce the PoP. Vendors can change memory configuration or graphics capability as needed.

2.5.4 SiP versus PoP

Comparing SiP and PoP, PoP has significant cost advantages over SiP because it essentially eliminates known-good-die issues and associated yield and procurement issues.

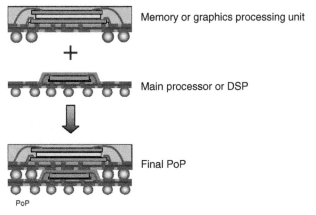

Figure 2.29: Example of a PoP

Source: www.freescale.com

PoP also enables a path using standard commodity memory footprints, rather than the custom designs typically required for SiP implementation (Table 2.3).

In contrast, PoP testing can be independent of the final SoC solution, so a PoP can use standard commodity memory design-for-test (DFT) modes. PoP also eliminates the procurement-related business issues of the SiP package, which is implemented by a logic chip supplier who must procure the memory chips from another IC supplier.

Table 2.3: Comparing SiP to PoP. Source: www.freescale.com

	Stacked CSP	Package-On-Package
Package/system cost	● X	△ △
Package reliability	●	△
Package mounting height	●	△
Package availability	●	△
Needs memory sub-package	●	X
PCB footprint	●	△
Customer SMT assembly	●	◉
Electrical performance	●	△
Thermal performance	◉	◉
Requires memory KGD	△	◉
QCT development resources	△	△
Memory test, repair, and burn-in	X	●
Flexible memory Configuration	X	●
Future memory standardization	X	△

Key:

Best - ● OK- △ Good- ◉ NG- X

The single impediment to PoP's proliferating, and taking over the SiP domain, is the lack of pinout and footprint standards similar to the industry-standard pinouts and package footprints established for most commodity memory products. Such standards would enable PoP manufacturers to leverage supply from commodity suppliers.

2.6 Summary

Batteries have only improved their capacity about 5% every 2 years [1]. As a result mobile devices need to become more power efficient to close the performance energy gap.

Success of power efficient mobile devices requires system-level optimization choices. However, a holistic approach consisting of systems design, software, architecture, circuit design, and manufacturing processes is required to tackle the low power challenges.

Further expansion in the specialization of device structures will proceed over the next few generations of CMOS, with increased emphasis on new materials and structures. This will maintain the momentum toward power, performance, and cost benefits that, until recently, had been simply benefits of scaling.

Further gains from scaling of traditional planar CMOS devices will be very difficult, limited by leakage and switching power considerations. The planar MOSFET will be challenged by multi-gate devices. Multi-gate devices offer an alternative path to increase the functions/unit silicon by providing better transistors for existing circuits and making new applications feasible using the novel features made possible by these devices.

As transistors get smaller, parasitic leakage currents and power dissipation become significant issues. By integrating the novel three-dimensional design of the tri-gate transistor with advanced semiconductor technology such as strain engineering and high-k/metal gate stack, an innovative approach has been developed toward addressing the current leakage problem while continuing to improve device performance.

The integrated CMOS tri-gate transistors will play a critical role in energy-efficient performance philosophy because they have a lower leakage current and consume less power than planar transistors.

Because tri-gate transistors greatly improve performance and energy efficiency, they enable manufacturers to extend the scaling of silicon transistors. Tri-gate transistors could become the basic building block for microprocessors in future technology nodes. The technology can be integrated into an economical, high-volume manufacturing process, leading to high-performance and low-power products.

The FinFET provides the most likely candidate for succession, enabling continued growth in density and reduction of cost for SoCs, even as the industry approaches nearing the atomic limit. The trends in benefits to density, performance, and power will be continued through such innovations. Rather than coming to a close, a new era of CMOS technology is just beginning.

The major challenges to sustain CMOS scaling include the economics and complexity of new materials and processes. Product innovation will be enhanced by process development and package contributions.

As mobile devices do more in less space, so do their enabling technologies. SoC packaging is no exception. Packaging technologies provide benefits not only to end device manufacturers, but also to consumers. Manufacturers gain the benefit of high levels of integration so that they can use increasingly sophisticated electronics without adding bulk. This integration also saves board space, so that more functionality, including cameras to multimedia players and radios, can be packed into mobile devices. Consumers also see the benefit of smaller mobile devices with more features, such as smartphones, PDAs, and portable media players.

While short-term packaging needs will be met by incremental improvements of current generations of technology, future packaging needs require new technology to meet evolving engineering and market demands. Each step along the path from SiP to PoP represents improvements in these two areas. Right now, each of these packages fit unique niches. For example, if size is most important, then stacked die will yield smaller packages. Moving into PoP may actually increase board space, but improves cost structure.

SoC, SiP, and PoP all provide varying degrees of functional and systems integration but are limited to conventional substrate design and manufacturing capabilities as well as traditional interconnect technologies such as wire bong and flip chip. SoC can become prohibitively expensive. However, the different waves of packaging techniques have helped produce decreases in system size that support Moore's law. Innovative packaging technologies like SiP and PoP are viewed as a critical enabler in helping continue this trend and fill what is called the **packaging gap**.

References

[1] C. Chun and A. Barth. eXtreme energy conservation for mobile communications. *Freescale Technology Forum*, July 2006.

[2] E.J. Nowak. Maintaining the benefits of CMOS scaling when scaling bogs down. *IBM Journal of Research and Development*, Vol. 46, No. 2/3, pp. 169–180, 2002.

[3] G.E. Moore. Cramming more components onto integrated circuits. *Electronics*, Vol. 38, Number 8, pp. April 18, 1965. Available at http://download.intel.com/museum/ Moores_Law/Articles Press_Releases/Gordon_Moore_1965_Article.pdf

[4] T. Skotnicki, J.A. Hutchby, T.J. King, H.S.P. Wong and F. Boeuf. The End of CMOS Scaling, *IEEE Circuits and Devices Magazine*, Vol. January/February, 2005, pp. 16–26.

[5] Z. Guo, S. Balasubramanium, R. Zlatanovici, T.-J. King and B. Nikolic. FinFET-based SRAM design. *Proceedings of the 2005 International Symposium on Low Power Electronics and Design*, August 2005, pp. 2–7.

[6] L. Mathew, Yang Du, S. Kaipat, M. Sadd, M. Zavala, T. Stephens, R. Mora, R. Rai, S. Becker, C. Parker, D. Sing, R. Shimer, J. Sanez, A.V.Y. Thean, L. Prabhu, M. Moosa,

B.Y. Nguyen, J. Mogah, G.O. Workman, A. Vandooren, Z. Shi, M.M. Chowdhury, W. Zhang, and J.G. Fossum, Multiple independent gate field effect transistor (MIGFET) – multi-fin RF mixer architecture, three independent gates (MIGFET-T) operation and temperature characteristics. *VLSI Technology, 2005. Digest of Technical Papers. 2005 Symposium on*, pp. 200–201, 14–16 June 2005.

[7] W. Zhang, J.G. Fossum and L. Mathew. Physical insights regarding design and performance of independent-gate FinFETs. *IEEE Transactions on Electron Devices*, Vol. 52, No. 10, 2198–2206, 2005.

[8] L. Mathew and J.G. Fossum. Freescale US Patent Application, February 2005.

[9] W. Zhang, J.G. Fossum, L. Mathew. IT FET: a novel FinFET-based hybrid device. *IEEE Transactions on Electron Devices*, Vol. 53, No. 9, 2006.

[10] B. Yu, L. Chang, S. Ahmed, H. Wang, S. Bell, C.-Y. Yang, C. Tabery, C. Ho, Q. Xiang, T.-J. King, J. Bokor, C. Hu, M-R. Lin and D. Kyser. FinFET scaling to 10 nm gate length. *Electron Devices Meeting, 2002. IEDM '02. Digest. International*, pp. 251–254, 2002.

[11] J. Kedzierski et al. *IEDM 2001*, pp. 437–440.

[12] B. Yu. Scaling towards 35 nm gate length CMOS. *Proceedings of the VSLI Symposium*, Kyoto, AMD, June 12–14, 2001, pp. 9–10.

[13] M. Jurczak, T. Skotnicki, M. Paoli, B. Tormen, J.-L. Regolini, C. Morin, A. Schiltz, J. Martins, R. Pantel and J. Galvier. SON (silicon on nothing) – a new device architecture for the ULSI era, *VLSI Technology, Digest of Technical Papers. Symposium on*, pp. 29–30, 1999.

[14] J. Hergentother. The vertical replacement-gate (VRG) MOSFET: A 50 nm vertical MOSFET with lithography-independent gate length, *IEDM 1999*, December 1999, p. 3.1.1, AT&T Bell Labs, p. 75.

[15] J.-T. Park and J.-P. Colinge, Multiple-gate SOI MOSFETs: device design guidelines, *Electron Devices, IEEE Transactions on*, Vol. 49, no. 12, 2222–2229, 2002.

[16] J.-J. Kim and K. Roy. *SOI Conference, 2003. IEEE International*, September 29–October 2, 2003, pp. 97–98.

[17] R. C. Phal and J. Adams. *Systems in Package Technology* (Presentation), *International Electronics Manufacturing Initiative*, SIP TIG Report, June 2005.

CHAPTER 3

Low Power Design Techniques, Design Methodology, and Tools

3.1 Low Power Design Techniques

Many design techniques have been developed to reduce power and by the judicious application of these techniques, systems are tuned for the best power/performance trade-offs.

3.1.1 Dynamic Process Temperature Compensation

A common engineering philosophy when designing a System-on-a-Chip (SoC) is to insure that they perform under "worst-case" conditions. Worst case in semiconductor manufacturing applies to very high temperatures and variations in the manufacturing process; transistor performance varies in a predefined range of parameters. Thus, some SoCs from the same wafer lot are capable of supporting higher operating frequencies (best case – fast process) or lower frequencies at the bottom of the predefined performance window (worst case – slow process) at the given voltage (Figure 3.1).

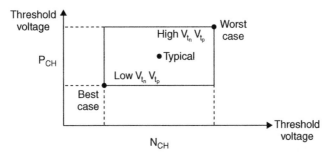

Figure 3.1: Process Variations Defining Varying SoC Performance

Dynamic process temperature compensation (DPTC) mechanism measures the frequency of a reference circuit that is dependent on the process speed and temperature. This reference circuit captures the product's speed dependency on the process technology and existing operating temperature and lowers the voltage to the minimum level needed to support the existing required operating frequency (Figure 3.2).

A mobile device containing a fast-process SoC operating in a moderate climate condition can be expected to work at the worst-case calculated voltage to support the required frequency. This is less than an optimum energy savings.

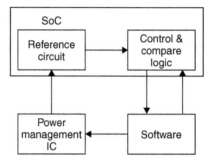

Figure 3.2: DPTC Mechanism

The DPTC concept allows the supply voltage to be adjusted to match the process corner and SoC temperature. If the process corner is "best case" a lower supply voltage can be applied to support the required performance of the SoC. Similarly, the temperature of the part can be used to adjust the supply voltage.

The available performance is monitored by different types of reference circuits comprised of free-running ring oscillators. The inputs from reference "sense" circuits are processed by internal control and compare logic and written to software readable registers. If there is a significant (predefined) change in the reference circuit delay values, an interrupt is triggered. The relevant software interrupt routine calculates the new required voltage and re-programs the Power Management IC (PMIC) to supply the new voltage to the SoC. A new voltage is applied, based on the reference circuit delay, values change, providing feedback and closing the loop of the DPTC mechanism. This insures that the system stabilizes at the proper voltage level. Software control permits fast and simple changes.

DPTC can result in an approximate power savings of 35%, significantly improving the battery life.

3.1.2 Static Process Compensation

Static process compensation (SPC) follows a similar path to DPTC but without the temperature compensation aspect. SoCs are designed at the worst-case process corner (see Figure 3.1). However, production wafer lots are typically manufactured close to a typical point in the "box" and as a result can run at a lower voltage and still meet performance requirements.

SPC is a technique of identifying minimum operating voltage for each SoC at the production line and programming the fuses with the information. Software reads the fuses to set the operating voltage for the SoC.

The basic circuits required to support SPC are similar to those employed in DPTC and integrated into the SoC. They include a ring oscillator, support register, and fuses (Figure 3.3).

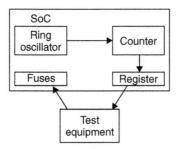

Figure 3.3: SPC Basic Circuits

The frequency of the ring oscillator correlates with the process corner. The support register captures the oscillator frequency and the fuses are programmed to define the counter frequency and voltage for the SoC.

3.1.2.1 Compared to DPTC

SPC does not compensate voltage for temperature and the operating voltage is not changed dynamically in SPC. In addition, the SoC manufacturer can test the SoC to the SPC defined voltage. Given that SPC is a subset of DPTC, it has been demonstrated that the temperature compensation aspect of DPTC provides marginal benefit to energy conservation.

3.1.3 Power Gating

Like voltage gating, power gating involves temporarily shutting down blocks in a design when the blocks are not in use. And, like voltage gating, the technique is complex. With power gating, the designer has to worry about it at the SoC design phase, specifically at the Register Transfer Level (RTL). The engineer has to design a power controller that is going to control what blocks need to shut down at a particular time and has to think about what voltage to run different blocks (Figure 3.4).

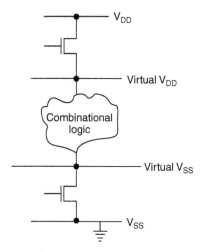

Figure 3.4: Power Gating

Traditionally, two methods for power gating are fine-grained and coarse-grained. In fine-grained power gating, designers place a switch transistor between ground and each gate. This approach allows designers to shut off the connection to ground whenever a series of functions is not in use. This technique is done with every cell in the library. There are trade-offs with fine-grained power gating because it is fairly easy to do power characterization of each cell. However, the problem is the area hit is very significant: two to four times larger (Figure 3.5).

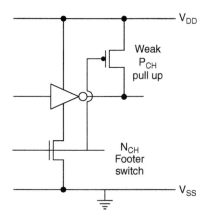

Figure 3.5: Fine Power Gating

In order to keep the area overhead to a minimum, fine-grained power gates are implemented as footer switches to ground as NMOS transistors. The timing impact of the IR drop across the switch and the behavior of the clamp are easy to characterize. It is still possible to use a traditional design flow to deploy fine-grained power gating.

Designers can also mix and match cells having some power gated and others not. Cells with high threshold voltage need not use power gating. For the most part, the power penalty is just too large, and many design groups are instead using coarse-grained power gating, in which designers create a power switch network. This is essentially, a group of switch transistors that in parallel turn entire blocks on and off. The technique does not have the area hit of the fine-grained technique because for a given block of logic the switching activity will be less than the 100%. However, due to the propagation delay through the cells, the switching activity will be distributed in time. In addition, it is harder to characterize on a cell-by-cell basis (Figure 3.6).

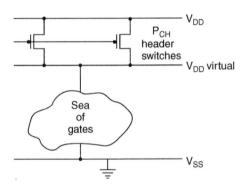

Figure 3.6: Coarse-Grained Power Gating

Unlike fine-grained power gating, when the power is switched in coarse-grained power gating, the power is disconnected from all logic, including the registers, resulting in the loss of all states. If the state is to be preserved when the power is disconnected then it must be stored somewhere, where it is not power gated. Most commonly this is done locally to the registers by swapping in special "retention" registers which have an extra storage node that is separately powered. There are a number of retention register designs which trade-off performance against area. Some use the existing slave latch as the storage node whilst others add an additional "balloon" latch storage node. However, they all require one or more extra control signals to save and restore the state.

The key advantage of retention registers is that they are simple to use and are very quick to save and restore state. This means that they have a relatively low energy cost of entering and leaving standby mode and so are often used to implement "light sleep." However, in order to minimize the leakage power of these retention registers during standby, it is important that the storage node and associated control signal buffering are implemented using high threshold low leakage transistors.

If very low standby leakage is required then it is possible to store the state in main memory and cut the power to all logic including the retention registers. However, this technique is more complex to implement and also takes much longer to save and restore state. This means that it has a higher energy cost of entering and leaving standby mode and so is more likely to be used to implement "deep sleep."

A key challenge in power gating is managing the in-rush current when the power is reconnected. This in-rush current must be carefully controlled in order to avoid excessive IR drop in the power network as this could result in the collapse of the main power supply and loss of the retained state.

3.1.4 State-Retention Power Gating

The major motivation of this technique is to significantly reduce the leakage power for the SoC when in the inactive mode. State-retention power gating (SRPG) is a technique that allows the voltage supply to be reduced to zero for the majority of a block's logic gates while maintaining the supply for the state elements of that block. The state of the SoC is always saved in the sequential components. Combinational elements propagate the state of the flip-flops. Using the SRPG technique, when in the inactive mode, power to the combinational logic is turned off and the sequential stays on. SRPG can thereby greatly reduce power consumption when the application is in stop mode, yet it still accommodates fast wake-up times.

Reducing the supply to zero in the stop mode allows both the dynamic and static power to be removed. Retaining the supply on the state elements allows a quick continuation of processing when exiting the stop mode.

Since the state of the digital logic is stored in the flip-flops, if the flip-flops are kept on a constantly powered voltage grid, the intermediate logic can be put onto a voltage grid that can be power gated. When the voltage is reapplied to the intermediate logic, the state of the flip-flops will be re-propagated through the logic and the system can start where it has left off as illustrated in Figure 3.7.

In a full SRPG implementation the entire target platform is entered into state retention and all (100%) flip-flops retain the state during power down. There is a specific power

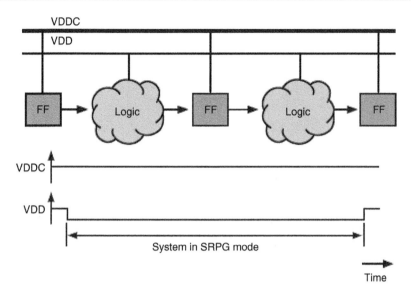

Figure 3.7: SRPG Technique

Source: http://www.freescale.com

up sequence required and the power down sequence is also predefined. Power up/down time is dependent on number of flops. Expected "wake-up" latency is less than 1 ns and "sleep" is less than 500 ns.

3.1.4.1 Partial SRPG

Partial SRPG further reduces the leakage from a full SRPG implementation.

In this case only a few flip-flops are made capable of state retention and all other flip-flops are turned off.

After power up, the SRPG flip-flops are restored to original state and the others are restored to reset state of that flip-flop. System software understands the non-saved registers are lost and should either re-program those registers or ensure the reset state meets the software requirements.

3.2 Low Power Architectural and Subsystem Techniques

3.2.1 Clock Gating

A tried-and-true technique for reducing power is clock gating. One-third to one-half of an IC design's dynamic power is in the SoC's clock-distribution system. The concept is simple, if you do not need a clock running, shut it down (Figure 3.8).

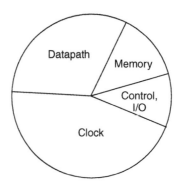

Figure 3.8: Power Distribution in a High Performance Processor [2]

Today, the two popular methods of clock gating are local and global. If you feed old data to the output of a flip-flop back into its input through a multiplexer, you typically need not clock again. Therefore, you can replace each feedback multiplexer with a clock gating cell that clocks the signal off. You would then use the enable signal that controls the multiplexer to control the clock cell to clock the signal off (Figure 3.9).

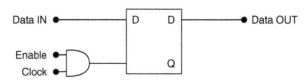

Figure 3.9: Local Clock Gating

The other popular approach of clock gating, global clock gating, is to simply turn off the clock to the whole block, typically from a central-clock-generator module. This method functionally shuts down the block, unlike local clock gating, but even further reduces dynamic power because it shuts down the entire clock tree (Figure 3.10).

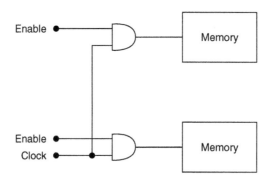

Figure 3.10: Global Clock Gating

3.2.2 Asynchronous Techniques: GALS

The increasing demands of mobile devices create several problems for system design and integration. Challenges like the integration of complex systems, timing closure including clock generation and control, system noise characteristic, and power consumption for mobile applications, in particular wireless devices.

Most digital systems designed today operate synchronously. One significant challenge is the generation of the clock tree. The clock tree in complex digital systems includes clock gating supporting circuitry, clock dividers for different clock domains, Phase Locked Loops (PLL), and complex clock-phasing blocks.

Mobile communication devices have one very critical constraint, power consumption. The limited capacity of the batteries creates firm limits for the system power consumption. In addition, the power demands of complex systems are usually high and hence, power consumption must be controlled and minimized. Partly, this can be achieved by using known methods for minimization and localization of switching power like clock gating, asynchronous design, or voltage scaling. However, clock gating makes the design of the clock tree even harder.

Clockless logic employs a protocol of local handshaking rather than a global clock to define transaction timing. Handshakes are implemented by means of simple request and acknowledge signals that mark the validity of the data. Only those parts of the system actively involved in the task execution draw power.

Globally asynchronous locally synchronous (GALS) techniques have the potential to solve some of the most challenging design issues of SoC integration of mobile devices.

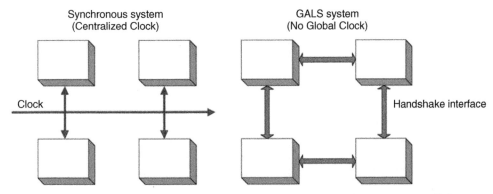

Figure 3.11: Comparison of a Synchronous Systems and GALS System [1]

The basic tenet behind GALS is that system blocks can internally operate synchronously and communicate asynchronously [1].

Figure 3.11 illustrates the two scenarios. The first is a normal synchronous system with a master clock, while the second is a GALS system in which two blocks talk to each other with a handshake interface. Although each block has its own local clock, the overall system works without a global clock. In SoC designs, the GALS architecture (Figure 3.12) helps to solve the increasingly difficult problem of integrating multiple clock domains into a single SoC and reducing the overall power consumption [2].

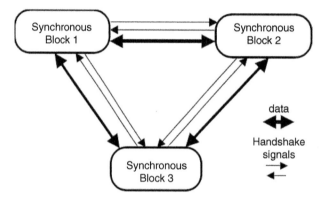

Figure 3.12: The GALS Architecture [2]

A GALS system consists of a number of locally synchronous modules each surrounded with an asynchronous wrapper. Communication between synchronous blocks is performed indirectly via asynchronous wrappers (Figure 3.13).

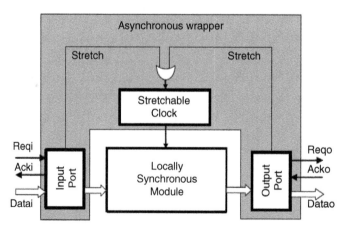

Figure 3.13: Communications Is Performed with Asynchronous Wrappers [2]

The purpose of the asynchronous wrapper is to perform all operations for safe data transfer between locally synchronous modules. Accordingly, a locally synchronous module should just acquire the input data delivered from the asynchronous wrapper, process it and provide the processed data to the output stage (port) of the asynchronous wrapper. Every locally synchronous module can operate independently, minimizing the problem of clock skew and clock tree generation.

A GALS design methodology should include power saving mechanisms. The goal is to completely integrate power saving mechanisms into the asynchronous wrappers. Implementing GALS would automatically introduce a certain power reduction. However, the power saving in GALS is based on the same assumptions as clock gating in the synchronous design. The main idea is identical, lowering of dynamic power consumption by disabling the clock signal.

Locally synchronous modules are usually surrounded by asynchronous wrappers. Local clocks drive those synchronous circuit blocks. Stoppable ring oscillators are frequently used to generate the local clocks. Data transfer between different blocks requires stopping of the local clocks during data transfer in order to avoid meta-stability problems. The asynchronous wrappers should perform all necessary activities for safe data transfer between the blocks. Locally synchronous modules do not play any role in providing the prerequisites for block-to-block data transfer.

The power savings of a GALS design has been investigated. Most of these investigations are based on the application of GALS to high-speed processor implementations. However, these results show some general trend. In Figure 3.14, the power distribution in a high performance processor is given.

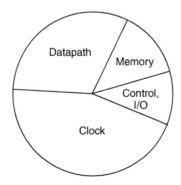

Figure 3.14: Power Distribution in a High Performance Processor [2]

The clock signal is the dominant source of power consumption in such a high performance processor. Within GALS design, the clock network is split into several smaller sub-networks with lower power consumption. First estimations, according to [11], showed that about 30% of power savings could be expected in the clock net due to the application of GALS techniques.

In addition, GALS design techniques offer independent setting of frequency and voltage levels for each locally synchronous module. When using dynamic voltage scaling (DVS), an average energy reduction of up to 30% can be reached, yet associated with a performance drop of 10%, as reported in [12]. The power saving techniques that are immanent with the use of GALS have similar limits as clock gating. Further energy reduction is possible only with the application of DVS in conjunction with GALS.

The ARM996HS processor is an example of a commercially available processor that uses Handshake Solutions clock less technology [3].

3.2.3 Power Saving Modes

SoCs offer many different power saving modes providing the mobile device developer the ability to trade-off between power consumption in standby and recovery times as shown in Figure 3.15:

> *Run*: This is the normal, functional operating mode of the SoC. Core frequency and operating voltage can be dynamically changed within a range.

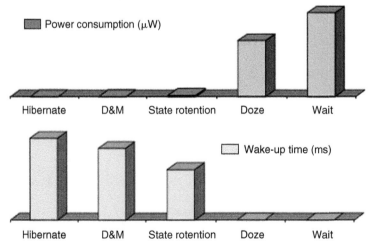

Figure 3.15: Power Saving Modes versus Wake-Up Time

Source: http://www.freescale.com

Wait: In this mode, the processor core clock is gated. Operation resumes on interrupt. This mode is useful for running low MIPS applications that primarily involve peripheral activity, such as a viewfinder.

Doze: In this mode, both the processor core and MAX clocks are gated. Clocks for specific peripherals can be switched off automatically in Doze mode by pre-programming the Clock Controller module. This mode is useful for processes that require quick reactivation. Normal operation resumes on interrupt.

State retention: In this mode, all clocks are switched off and the PLL is disabled. External memory is put in low power (self-refresh) mode. MCU and peripheral clocks are gated. Supply voltage can be dropped to a minimum. State-retention uses less power and has a longer wake-up time than Doze mode, but there is no need to recover any data after the wake-up.

Deep Sleep: In this mode the clocks are gated. MCU core platform power supply is turned off. Any relevant register data should be saved before entering Deep Sleep mode. Normal operation resumes on interrupt.

Hibernate: The power supply of the entire SoC is shut down. System is completely dead. Operation resume is equivalent to a cold boot. All internal data should be saved to external memory prior to hibernate.

The power saving modes can be extended to board-level applications. For example, in a memory hold mode, everything can be powered down except the PMIC and memory, which is kept in a state-retention mode. This is a very low power state, and in many applications different components, such as an application or baseband processor, will have to reboot at wake-up, creating some latency issues.

However, if the core processor is kept in a low leakage standby mode, wake-up is much quicker because the baseband reboot will not be necessary. The application requirements will often dictate which memories hold state is used for best performance/energy efficiency optimization.

3.3 Low Power SoC Design Methodology, Tools, and Standards

3.3.1 Introduction

There has been increasing concern in system design houses and semiconductor suppliers that power consumption of ICs is beginning to impact design complexity, design speed, design area, and manufacturing costs.

Historically power consumption has not been an issue or a high priority in the semiconductor industry. The industry was driven by the competitive market forces of simultaneously increasing integration and reducing SoC die area. In the high-volume processor market segment, integration was driven by the need to increase the performance. Higher performance was achieved through the use of new architectures (e.g. super scalar and super pipelined); integrating more and more system functions on the SoC increasing the processor's speed and increasing the processor's ability to perform computations. The trend in the large, and cost sensitive, embedded processor market segment is toward higher frequencies.

Today one in four IC designs fail due to power-related issues. Power management techniques are considered a critical part of the SoC design flow.

In rising to the challenge to reduce power the semiconductor industry has adopted a multifaceted approach, attacking the problem on three fronts:

1. *Reducing chip capacitance through process scaling*: This approach to reduce the power is very expensive and has a very low return on investment (ROI).

2. *Reducing voltage*: While this approach provides the maximum relative reduction in power it is complex, difficult to achieve and requires moving the DRAM and systems industries to a new voltage standard. The ROI is low.

3. *Employing better architectural and circuit design techniques*: This approach promises to be the most successful because the investment to reduce power by design is relatively small in comparison to the other two approaches.

The need to reduce power consumption has become more critical as larger, faster ICs move into portable applications. As a result techniques for managing power throughout the design flow are evolving to assure that all parts of the product receive power properly and efficiently and that the product is reliable. However, there is one significant drawback, namely the design problem has essentially changed from an optimization of the design in two dimensions, i.e. performance and area, to optimizing a three-dimensional problem, i.e., performance, area, and power (see Figure 3.16).

As stated earlier, the best power management decisions are made at the system level during system design. At this point in the development cycle, system designers make initial decisions about the end-product power requirements and distribution. That entails determining the overall power budget, making critical hardware/software trade-offs, and working with designers to select the proper process technology for semiconductor

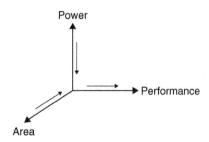

Figure 3.16: Performance, Area, and Power Trade-Off

fabrication, such as gate length and multi-voltage/multi-threshold. Electronic Design Automation (EDA) gives system designers the design and verification environment they need to achieve power/performance estimation at the system level. Such system-level decisions drive all subsequent software and semiconductor requirements.

Just as system-level decisions drive the semiconductor requirements, ICs together with process technology drive EDA tool and flow requirements. IC designers select EDA tools and flows for power management based on the following criteria:

- Process and application support.

- Design optimization capabilities – needed to ensure operations stay within power budget.

- Design verification and signoff capabilities – needed to ensure device reliability and power integrity.

EDA tools address all these requirements, offering power management throughout the design flow. The most effective EDA power management solutions are characterized by the ability to deliver comprehensive power estimation, optimization, analysis, and reliability signoff within a power-aware synthesis and signoff environment.

Unlike timing, which is localized to critical paths within the IC, power is distributed across the device. Power requirements are also dependent on the particular application software that is running and the operating mode. In general, three areas of power need to be managed in the flow to ensure the design remains within the required power budget and reliability targets:

- *Dynamic power*: The designer must manage how much power is used when a circuit switches or is active (changes from 1 to 0 or 0 to 1) due to charging/discharging capacitance.

- *Leakage (static) power*: The designer must manage how much power "leaks" from transistors or is wasted when the circuit is static or unchanged.

- *Peak power*: The designer must manage peak-load periods.

A variety of techniques have been developed to address the various aspects of the power problem and to meet aggressive power specifications. Examples of these include the use of clock gating, multi-switching threshold transistors, multi-supply multi-voltage, substrate biasing, dynamic voltage and frequency scaling, and power shut-off (PSO). Table 3.1 illustrates the power, timing, and area trade-offs among the various power management techniques.

Table 3.1: Cadence Low Power Solution Architecting, Designing, Implementing, and Verifying Low Power Digital SoCs. www.cadence.com

Power Reduction Technique	Power Benefit	Timing Penalty	Area Penalty	Methodology Impact			
				Architecture	Design	Verification	Implementation
Multi-Vt optimization	Medium	Little	Little	Low	Low	None	Low
Clock gating	Medium	Little	Little	Low	Low	None	Low
Multi-supply voltage	Large	Some	Little	High	Medium	Low	Medium
Power shut-off	Huge	Some	Some	High	High	High	High
Dynamic and adaptive voltage frequency scaling	Large	Some	Some	High	High	High	High
Substrate biasing	Large	Some	Some	Medium	None	None	High

The use of more advanced approaches reduces power consumption, but at the same time increase the complexity associated with design, verification, and implementation methodologies. Although using a single technique in isolation could be relatively simple, often a combination of these techniques must be used to meet the required timing and power targets. Using multiple techniques concurrently could result in an extremely complex design flow. A key requirement is consistency throughout the flow such that the use of one technique preserves any gains from other techniques. This also requires a good understanding of the various low power techniques and their respective trade-offs, as highlighted in Table 3.1.

Furthermore, "low power" is not just something that is "bolted" on at the end of the development process. To meet aggressive design schedules, it is no longer sufficient to consider power only in the implementation phase of the design. The size and complexity of today's ICs makes it imperative to consider power throughout the design process, from the SoC/system architectural phase; through the implementation architecture phase; through design (including micro-architecture decisions); and all the way to implementation with power-aware synthesis, placement, and routing. Similarly, to prevent functional issues from surfacing in the final silicon, power-aware verification must be performed throughout the development process.

3.3.2 Low Power Design Process

3.3.2.1 SoC Design Flow

To address these issues, power management techniques should be applied during every step of the flow, from design planning through reliability signoff analysis (see Figure 3.17).

3.3.2.2 System Design

For a new product, a well-defined product specification is the first critical step to achieving a successful product. The design specification typically is driven by a product requirements document (PRD), or user specification, typically generated by the technical marketing organization, which contains a synthesis of customer requirements in the form of features, cost, and schedule.

The design or architectural specification is derived from the PRD. The architectural specification can be hierarchical in nature with the top level describing an abstract picture of the SoC under development. The subsequent layers divide the top-level blocks into finer detail and at the lowest layer the SoC designer is left with a description of all the IP blocks that comprise the SoC.

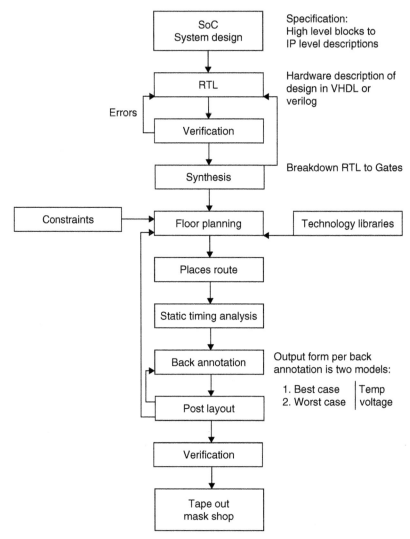

Figure 3.17: SoC Design Flow

3.3.2.3 RTL Design

In a SoC design, RTL description is a way of describing the operation of a digital circuit. In RTL design, a circuit's behavior is defined in terms of the flow of signals (or transfer of data) between registers, and the logical operations performed on those signals.

This step converts the user specification (what the user wants the SoC to do) into a RTL description. The RTL specifies, in painstaking detail, exactly what every bit of the SoC should do on every clock cycle.

RTL abstraction is used in hardware description languages (HDLs) like Verilog and VHDL to create high-level representations of a circuit, from which lower-level representations and ultimately actual wiring can be derived.

When designing digital ICs with a HDL, the designs are usually engineered at a higher level of abstraction than transistor or gate level. In HDLs, the designer declares the registers (which roughly corresponds to variables in computer programming languages), and describes the combination logic by using constructs that are familiar from programming languages such as if-then-else and arithmetic operations. This level is called Register Transfer Level. The term refers to the fact that RTL focuses on describing the flow of signals between registers.

3.3.2.4 Verification

Verification is the act of proving or disproving the correctness of intended algorithms underlying a system with respect to a certain formal specification. It answers the question "are we building the product correctly?"

3.3.2.5 Synthesis

Using an EDA tool for synthesis, the RTL description can be directly translated to equivalent hardware implementation files, typically in terms of logic gates, for an SoC. Logic gates can take the form of AND, OR, NAND, NOR, and other logical functions.

During synthesis, EDA tools help the designer complete RTL and gate-level power optimization together with peak and average power debug and analysis. Here, power management is needed to ensure the design's dynamic and leakage power remains within the power budget and reliability targets. Power optimization and peak power analysis solutions include support for multi-voltage, multi-threshold, clock gating, data, and power-gating; RTL estimation analysis at full-SoC/block level; and gate-level peak and average power analysis. The synthesis tool also performs logic optimization.

3.3.2.6 Floorplanning

During Floorplanning, the RTL of the SoC is assigned to gross regions of the SoC, input/output (I/O) pins are assigned and large objects (arrays, cores, etc.) are placed.

From a low power perspective EDA tools help designers create the power grid and perform power network analysis to assure the power remains within the power budget and reliability targets. Power-aware floorplanning solutions include support for power grid planning, multi-voltage region planning, clock tree planning, and estimation of voltage (IR) drop and electromigration effects.

3.3.2.7 Place and Route: Physical Implementation

During Place and Route, a placer takes a given synthesized circuit, an output from the Floorplanning, together with a technology library and produces a valid placement layout. The layout is optimized according to the constraints and ready for cell resizing and buffering, considered an essential step for timing and signal integrity satisfaction. Clock tree synthesis (CTS) and routing follow, completing the physical design process.

Routing is the process of creating all the wires needed to properly connect all the placed components, while obeying the design rules of the process. Place and Route are often lumped together as necessary physical operations.

During Place and Route, EDA helps designers implement power saving features to manage both dynamic and leakage power in the design. Power implementation solutions today include support for multi-voltage and multi-threshold designs.

3.3.2.8 Tape-out and Mask Making

In SoC design, tape-out is the final stage of the design phase of an SoC such as a microprocessor. It is the point at which the description of a circuit is sent for manufacture, initially to a mask shop and then to a wafer fabrication facility.

3.3.2.9 Reliability Signoff Analysis

During reliability signoff analysis, EDA provides post-route analysis of power grids to detect power anomalies such as voltage drop and electromigration which, if left uncorrected, can lead to design failure.

Power management starts at the system level and is driven by the application power budget requirements throughout the semiconductor value chain. IC designers collaborate closely with their EDA tool providers to ensure power management is an integral part of their design flow.

EDA offers designers the much-needed infrastructure and tools for enabling power management throughout the flow. There are future opportunities for EDA to contribute to more power savings during system definition. The design for low power problem cannot be achieved without good tools. The following sections will cover the design process, some key EDA vendor approaches and standardization.

3.3.3 Key EDA Vendors Approach to Low Power Design

3.3.3.1 Cadence Low Power Flow

Cadence has developed a complete solution for the design, verification, and implementation of low power SoCs. The low power solution combines a number of technologies from several Cadence platforms. Each of these advanced products also leverages the Common Power Format (CPF) (Figure 3.18).

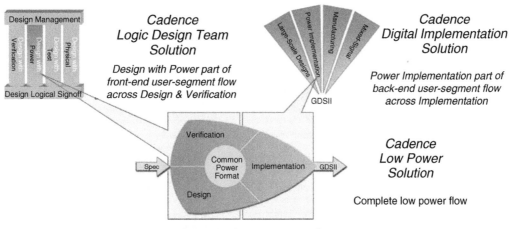

Figure 3.18: Low Power Flow

Source: http://www.cadence.com

3.3.3.2 Cadence Low Power Methodology Kit

Low Power Methodology Kit, offered by Cadence, provides users with a complete end-to-end methodology for low power implementation and verification [4] (Figure 3.19).

The kit is modularized to allow for incremental adoption and to allow teams to focus on what is most critical for the design (Figure 3.20).

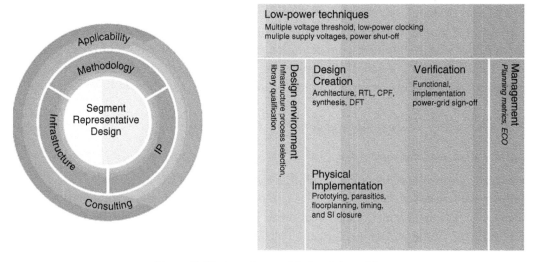

Figure 3.19: Low Power Methodology Kit

Source: http://www.cadence.com

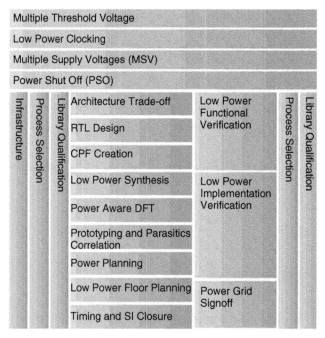

Figure 3.20: Modular Kit

Source: http://www.cadence.com

The Cadence Low Power Methodology Kit includes the following:

- Wireless segment representative design (SRD), including all required views for RTL design, physical implementation, and verification (including test bench)

- Detailed low power methodology guide, covering all aspects of low power implementation

- Reference flow implementations with step-by-step walkthroughs

- Detailed documentation of the SRD and reference flow

- Detailed flow checklists and trade-off analysis

- Expert consulting designed to map the verified and demonstrated methodologies to a specific customer design

The Low Power Methodology Kit utilizes and integrates with the following technologies (Figure 3.21):

- Cadence Logic Design Team Solution

- Incisive Plan-To-Closure Methodology

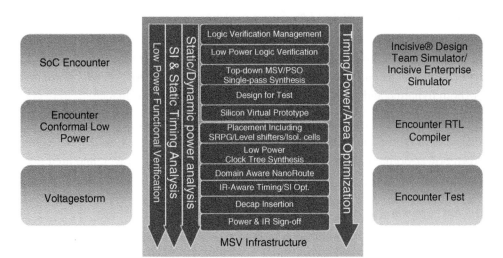

Figure 3.21: Low Power Design Flow

Source: http://www.cadence.com

- Incisive Enterprise Family

- Incisive Formal Verifier

- Encounter RTL Compiler

- Encounter Test

- First Encounter

- SoC Encounter

- Encounter Conformal EC

- Encounter Conformal LP

- Voltage Storm

3.3.3.3 Incisive Design Team and Enterprise Simulators

The Incisive simulators help to capture functional failures due to inserted power management structures and identify bugs that could lead to excessive power consumption. In addition, the incisive simulators help engineers explore different power architectures early in the design flow without having to wait for gate-level netlists or needing to develop resource consuming modeling techniques.

3.3.3.4 Encounter RTL Compiler: Low Power Synthesis

It performs multi-objective optimization that simultaneously considers timing, power, and area to create logic structures that meet all these goals in a single pass. Encounter RTL Compiler synthesis techniques address advanced SoC design needs, such as decreased power (both dynamic and static leakage) without sacrificing performance.

Encounter RTL Compiler global synthesis allows designers to use the CPF to explore the trade-offs of employing different power management techniques. It then performs top-down power-domain-aware synthesis, optimizing for timing, power, and area simultaneously.

3.3.3.5 SoC Encounter

The SoC Encounter system combines RTL synthesis, silicon virtual prototyping, and full-SoC implementation in a single system. It enables engineers to synthesize to a flat virtual prototype implementation at the beginning of the design cycle. With the SoC Encounter system, engineers have a view of whether the design will meet its targets and be physically realizable. Designers can then choose to either complete the final implementation or to revisit the RTL design phase. It also supports advanced timing closure and routing, as well as signoff analysis engines for final implementation.

3.3.3.6 Encounter Conformal LP

During development, a low power design undergoes numerous iterations prior to final layout, and each step in this process has the potential to introduce logical bugs. Encounter Conformal Low Power checks the functional equivalence of different versions of a low power design at these various stages and enables the identification and correction of errors as soon as they are introduced. It supports advanced dynamic and static power synthesis optimizations such as clock gating and signal gating, Multi-Vt libraries, as well as de-cloning and re-cloning of gated clocks during CTS and optimization.

In addition, it supports the CPF specification language. It uses CPF as guidance to independently insert and connect low power cells (level shifters, isolation, and state-retention registers) into an RTL design, thus enabling true low power RTL-to-gate equivalence checking.

3.3.3.7 Encounter Test

Based on the power intent read through the CPF, Encounter Test automatically creates distinct test modes for the required power domain and PSO combinations. It inserts special purpose design-for-test (DFT) structures to enable control of PSO during test. The power-aware automatic test pattern generation (ATPG) engine targets low power structures such as level shifters and isolation cells, and generates low power scan vectors that significantly reduce power consumption during test. This results in the highest quality of test and the lowest power consumption during test for low power devices.

3.3.3.8 VoltageStorm

VoltageStorm hierarchical solution gives design teams the confidence that IR (voltage) drop and power rail electromigration are managed effectively. VoltageStorm technology has evolved to become an integral component of design creation, which requires early and up-front power rail analysis to help create robust power networks during power planning. Employing parasitic extraction that is manufacturing aware, and using static and dynamic algorithms, VoltageStorm technology delivers power estimation and power rail analysis functionality and automation to both analyze and optimize power networks throughout the design flow.

3.3.3.9 Synopsis

Synopsis provides an automated, end-to-end power management solution that delivers the low power implementations. It ensures consistent correlation from RTL to silicon, enabling design teams to benefit from reduced iterations and improved productivity. The Synopsys' solution utilizes industry standards such as Unified Power Format (UPF) and is complemented by a strong ecosystem of IP, modeling techniques, and libraries (Figure 3.22).

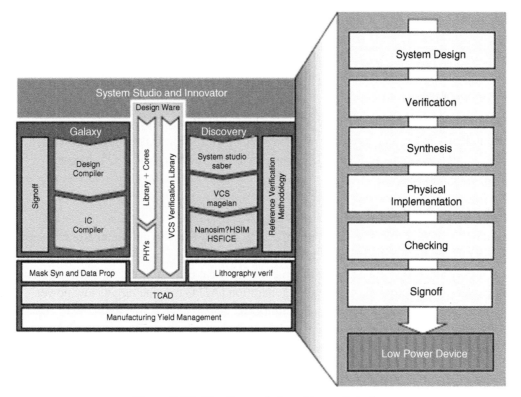

Figure 3.22: The Synopsis Low Power Solution

Source: http://www.synopsis.com

3.3.3.10 System Design

Early insight into power consumption at the system level enables the most significant power savings. Synopsys' **DesignWare® Virtual Platforms** [5] enable power modeling at the system level, providing a relative measure of power consumption on an application-by-application basis using the real applications' software and target OS.

Virtual platforms effectively contribute to lowering the risk and decreasing the time-to-market, through early software development and hardware/software integration. The benefits of virtual platforms include:

- Enabling concurrent hardware/software development, by using Virtual Platform technology and models, system and application software development can begin months earlier in the product development cycle, and concurrent with hardware development.

- Reducing the SoC-based product release cycle by 6–9 months.

- Improving software development productivity, typically by a factor of 2-5x, by allowing unlimited observability and controllability of the target hardware, and by predictable and repeatable execution of debug scenarios.

- Promoting continuous hardware/software integration, rather than software and hardware meeting near the end of silicon development which is no longer practical in today's fast-paced markets.

The **Innovator** integrated development environment (IDE) helps designers develop, run and debug Virtual Platforms. Innovator allows designers to rapidly build Virtual Platforms and instantiate, configure, and connect low power components for early and accurate exploration of low power architecture.

System Studio is a system-level design creation, simulation, and analysis tool that supports transaction-level modeling. System Studio allows design teams to make early hardware–software architectural trade-offs, including modeling of power management strategies.

3.3.3.11 Verification

VCS® solution is the most comprehensive RTL verification solution in a single product, providing advanced bug-finding technologies, a built-in debug and visualization environment and support for all popular design and verification languages including Verilog, VHDL, SystemVerilog, and SystemC. The VCS solution understands and acts on the power intent definitions, enabling correct simulation of power domains in all modes of operation.

Verification tools use the definition of power domains and how they are controlled and shut down during RTL simulation to verify the functional behavior. Similarly, simulation must be able to model the use of isolation and retention cells so that designers can model the functionality correctly in different power states.

3.3.3.12 RTL Synthesis

The design team should know as early as possible if there is any risk that a design is not going to meet its power budget. In addition, the RTL synthesis should exploit all available power optimization techniques concurrently with timing optimization and all of the other optimization parameters.

Power cannot be optimized in isolation, it must be done in the context of area, speed, yield, manufacturability, reliability, signal integrity, as well as for worst-case process corners/variability and low power/voltage scenarios.

Design Compiler® **Ultra** with topographical technology delivers accurate correlation to post-layout power, timing, test, and area, without wire load models by using the same placement as layout engine. It is designed for RTL designers and requires no physical design expertise or changes to the synthesis use model. The accurate prediction of layout power, timing, area, and testability in Synopsys' Design Compiler Ultra enables RTL designers to identify and fix real design issues while still in synthesis and generate a better start point for physical design. The clear design benefit is reduced iterations and faster turnaround time.

Power Compiler automatically minimizes dynamic power consumption at the RTL and gate level, using both clock gating and simultaneous optimization for power, timing, and area. Power Compiler also supports the use of multi-threshold libraries for automatic leakage power optimization.

Power Compiler performs automatic clock gating at the RTL without requiring any changes to the RTL source. This enables fast and easy trade-off analysis and maintains technology-independent RTL source.

DFT Compiler enables designers to conduct in-depth testability analysis of RTL, to implement the most effective test structures at the hierarchical block level, and, if necessary, to automatically repair test design rule checking violations at the gate level (Figure 3.23).

DFT Compiler transfers all information about the Adaptive Scan architecture to TetraMAX to automatically generate compressed test patterns with the highest test coverage.

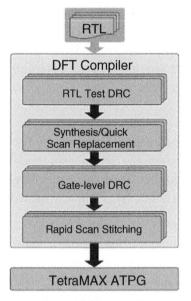

Figure 3.23: DFT Compiler Flow

Source: http://www.synopsis.com

TetraMAX ATPG automatically generates high-quality manufacturing test vectors. It is the only ATPG solution optimized for a wide range of test methodologies that is integrated with Synopsys' test synthesis tool. TetraMAX allows RTL designers to quickly create efficient, compact tests for even the most complex designs.

3.3.3.13 Physical Implementation

The Place and Route solution, **IC Compiler**, performs power-aware placement. By shortening high-activity nets and clustering registers close to the clock source. The clock tree has a critical part to play in dynamic power dissipation. Power-aware CTS can take advantage of merging and splitting integrated clock gating cells (ICGs) in the placed design. This capability allows more efficient use of ICGs, which saves significant dynamic power in the clock tree.

IC Compiler is a complete physical implementation solution with hierarchical design planning with automated power network synthesis and analysis. IC Compiler includes Extended Physical Synthesis (XPS) technology which provides a single convergent flow from netlist to silicon with support for multi-voltage designs, multi-threshold leakage, low power placement and CTS, and state-retention power gating.

JupiterXT™ design planning solution enables fast feasibility analysis for a preview of implementation results, and provides detailed floorplanning capabilities for flat or hierarchical physical design implementation styles. In addition, it provides fast, automatic placement, power network synthesis, and in-place optimization to allow designers to quickly generate prototype floorplans.

3.3.3.14 Checking

In multi-voltage designs, voltage-level shifting cells, and power domain isolation cells affect signal functionality; therefore, the design team needs to specify power intent in the RTL code to support automated RTL-to-gate equivalence checking. To ensure design integrity, designers also need to check special power-related structures against a whole battery of design rules.

Formality is used to make sure the RTL is equivalent to the gate-level implementation of the multi-voltage design. **Leda** checker is a programmable design and coding guideline checker that can be used to validate an entire SoC against a full range of low power checks. Leda provides over 50 low power checks, including insertion/location of level shifters and isolation cells and clock gating used to turn off power regions.

3.3.3.15 Signoff

Signoff must include power. Power management decrees whether an SoC will meet specifications. Power grid IR drop and multi-voltage design techniques directly affect SoC-level timing, so timing signoff must include the impact of voltage and temperature variation.

PrimeRail delivers full-SoC dynamic voltage-drop and electromigration analysis for power network signoff. By ensuring accurate correlation with design planning designers can be confident that the important decisions they take during the design planning stages with respect to power network planning will not have to be revisited at the end of the design cycle.

PrimeTime® **PX** enables full-SoC, concurrent timing, signal integrity, and power analysis in a single, correlated environment. It provides high accuracy dynamic and leakage power analysis concurrently with timing and signal integrity analysis. This capability offers major accuracy and productivity benefits over the use of separate or standalone timing and power analysis tools.

HSIMplus™ PWRA provides power net reliability analysis of IR drop and electromigration in power and ground buses. Concurrently **HSIMplus™ PWRA**

accounts for the effects of interconnect resistance on dynamic circuit performance. In addition, it overcomes the limitations of simulation with millions of extracted parasitic resistors by incorporating built-in compression and reduction algorithms to maintain accuracy, capacity, and performance.

3.3.4 Low Power Format Standards

Limitations in the existing design infrastructure prevent power-related aspects of design intent from being specified across the design chain. The industry lacks automated power-control techniques capable of achieving both functional and structural verification of designs prior to incurring extensive manufacturing costs. Threatened by potentially high costs coupled with missed time-to-market opportunities, companies will remain reluctant to adopt advanced process geometries and effective low power methodologies with advanced process technologies.

The reluctance to adopt new process geometries has a negative impact across all business sectors in the electronics industry. Innovation by design teams is constrained due to the risk of low yields and costly re-spins. Library and IP vendors are unable to leverage the differentiated value of modeling the new processes, and tool providers are unable to sell new capabilities based on evolving process requirements. As a result, consumers are offered products that suffer from shorter battery life, higher heat dissipation, and other shortcomings that lead directly to lower sales – negatively affecting the profitability of businesses across the industry.

A key enabler of a modern power-aware design flow is the ability to capture and preserve the intent of the SoC architects and designers throughout the design flow. This requires a common specification that can be used and shared across the entire design chain, from architectural specification to verification. Power management concept is not solved for EDA tools and silicon libraries exhaustively.

Immature and proprietary power management tools have been developed by many EDA companies. Historically, there has not been any dialog in the industry to develop a power management SoC tool flow. There are point solutions but no complete solution (Figure 3.24).

Currently, EDA tools debugging capabilities are poor due to complexity issues and weak and misleading reports. In addition, functional correctness is difficult to verify. Another concern for the SoC designer is that most power management tools are at the gate level rather than at RTL level to meet the accuracy requirements.

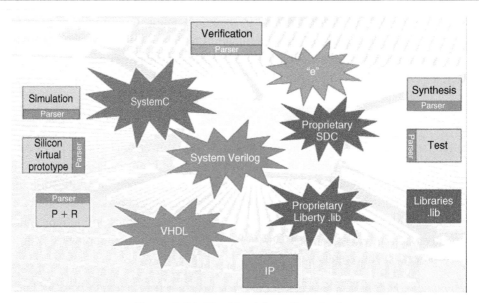

Figure 3.24: The Need for a Standard

Source: http://www.cadence.com

Power management problems concern all elements in the development chain, from system houses, silicon, IP providers, and tool providers that provides a 360 degree point of view.

Standardization is a key in building an efficient development flow and methodology. It insures a common language, a set of general methodologies developed around the standard, interoperable tools, and an opportunity for invention and creativity.

Power management tool requirements from SoC developers include:

Usability: Fills in existing tool flows minimizing learning curves, set up and deployment time.

Portability: Non-conflicting IP- and SoC-level activities. Insuring IP deliverables include information on how to implement and verify Power Management for IP when integrated into the SoC.

Reusability: Required for derivative designs.

Interoperability between tools: No scripting required to have tools work together.

Automatic verification: First time verification of the power management implementation accomplished automatically.

Standards should be speedily authored and employ open standards for low power design sourced by a broad inclusive team from across the SoC design chain.

Speed: Start simple, address immediate need, and evolve over time. Fill in the gaps by making use of existing standards and technologies.

Open: Everything on the table. No IP licensing or confidentiality required.

Inclusive: Invite everyone who wants an industry-wide solution to participate.

3.3.4.1 Single File Format

Currently, there are two major efforts in developing a common power description format that could be read by tools throughout the design flow [6]:

1. Common Power Format
2. Unified Power Format

3.3.4.2 Common Power Format

An example of such a specification is the CPF [7], which is managed by the **Silicon Integration Initiative** (Si2) consortium's **Low Power Coalition**. CPF is a design language that addresses limitations in the design automation tool flow. It provides a

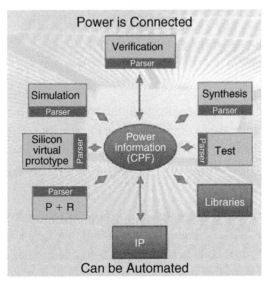

Figure 3.25: Standards-Based Solution

Source: http://www.cadence.com

mechanism to capture architects' and designers' intent for power management and it enables the automation of advanced low power design techniques. CPF allows all design, implementation, verification, and technology-related power objectives to be captured in a single file and then applies that data across the design flow, providing a consistent reference point for design development and production (Figure 3.25).

There are three major benefits of using CPF to drive the design, verification, and implementation steps of the development flow:

1. It helps designers achieve the required SoC specs by driving the implementation tools to achieve superior trade-off among timing, power, and area.

2. By integrating and automating the design flow, it increases designer productivity and improves the cycle time.

3. By eliminating the need for manual intervention and replacing ad hoc verification methodologies, it reduces the risk of silicon failure due to inadequate functional or structural verification.

3.3.4.3 Power Forward Initiative

The Power Forward Initiative's primary goal is to remove the barriers to automation of advanced low power design, and to provide a pathway to the development of a standards-based solution. By reducing the risks and costs associated with the design, verification, and implementation of designs using multiple/variable supply voltages and PSO design techniques, particularly at 90 nm and below, the door to a new era of innovation be opened.

Building a solution that fills the current void in the industry's infrastructure requires a departure from current approaches that have failed to create a holistic solution. In an effort to create a reset, the Power Forward Initiative addresses the entire scope of the industry – from the initial specification of design intent through to manufacturing and test (Figure 3.26).

However, this reset must not be disruptive to the portions of the existing infrastructure that are clearly working. What is required is a solution that serves as an overlay instead of a radical departure – one that does not require re-engineering of libraries, IP blocks, verification suites, or designs that can be reused. By definition, there are additive elements to the design environment and design process, but these new elements must not create a snowballing change throughout the ecosystem.

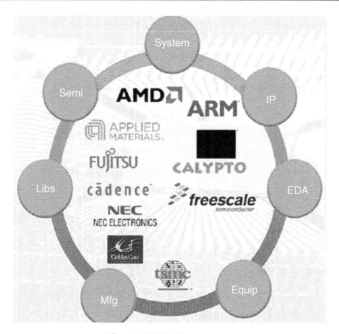

Figure 3.26: Ecosystem

Source: http://www.cadence.com

The Power Forward Initiative's initially focus is on the issues related to the use of complex power domains including multiple/variable supply voltages and power shut-off (PSO) techniques.

To achieve these goals, ongoing work on the **Common Power Format** is being considered within the larger industry frame of reference. Looking to satisfy the needs of the broad constituency of SoC design teams and providers of tools, equipment, IP, silicon, manufacturing, test, and services, the CPF has been created to deliver a comprehensive solution to the challenges posed by today's advanced power requirements. The CPF is being architected for future support of new design techniques and materials breakthroughs, including architecture, hardware and software system modeling, as well as analog and mixed-signal design.

Benefits of the CPF-based solution include:

- Functional verification of PSO using the same design description used for implementation

- Reduced turnaround time for physical implementation

- Enhanced optimization through simplified design exploration

- More accurate power utilization estimates

- Equivalence checking between functional description and implementation

Due to the fact that the new Power Forward approach is additive to the existing environment infrastructure and design flows, no changes to existing/legacy designs will be required. The CPF shows strong indications of viability for delivering productivity gains and improved Quality-of-Service (QoS) across the design chain. Automation is a key aspect to solve the industry's power management design challenges, and support for the CPF's extendable infrastructure will foster the adoption of automated low power design and verification solutions.

3.3.4.4 Unified Power Format

The UPF standard is a convergence of proven technology donations from a number of vendors [8]. EDA vendor contributions are derived from several years of successful use of their products on taped out low power designs. End customers contribute their internally developed optimization and analysis technologies which deal with application-specific power issues, especially for wireless and handheld devices. Strong collaborative participation by Accellera members and other dedicated companies results in an open standard.

When power consumption is a key consideration, describing low power design intent with the UPF improves the way complex ICs can be designed, verified, and implemented. The open standard permits all EDA tool providers to implement advanced tool features that enable the design of modern low power ICs. Starting at the RTL and progressing into the detailed levels of implementation and verification, UPF facilitates an interoperable, multi-vendor tool flow and ensures consistency throughout the design process.

A UPF specification defines how to create a supply network to supply power to each design element, how the individual supply nets behave with respect to one another, and how the logic functionality is extended to support dynamic power switching to these logic design elements. By controlling the operating voltages of each supply net and whether the supply nets (and their connected design elements) are turned on or off, the supply network only provides power at the level the functional areas of the SoC need to complete the computational task in a timely manner.

3.3.4.5 Accellera

Accellera, similar to Si2, provides design and verification standards for quick availability and use in the electronics industry. The organization and its members cooperatively deliver much-needed EDA standards that lower the cost of designing commercial IC and EDA products [9].

3.3.4.6 IEEE

In the near future SoC designers will have to cope with two standard power formats. Technically CPF and UPF are similar up to 90%. Both UPF and CPF use commands allow user to establish and manage separate power domains, set up level shifters, specify isolation and retention, and define power-related rules and constraints.

However, there is a silver lining, the IEEE has a working group currently studying both the CPF and UPF standards [10].

Clearly, this is one of the biggest challenges the EDA standards community has faced. An aggressive timetable, multiple standards organizations, and a highly competitive landscape make the task appear daunting.

3.4 Summary

Low power technologies and techniques for energy efficiency are distinctive combinations of advanced architectural techniques and circuit design techniques with the latest design methodology. These technologies and techniques provide the highest possible performance levels within a restricted power budget. They can be applied to any application, benefiting the user by extending battery life without impacting performance as well as keeping energy costs lower.

Various low power circuit and architectural techniques, for mitigating leakage power, have been described in this chapter. These include power gating, dynamic process temperature control, static process compensation, state-retention power gating, clock gating, low power modes, and asynchronous circuits.

Power consumption of mobile devices has moved to the forefront of SoC development concerns. Designing for low power can have a significant impact on the power budget of a mobile device. In addition, identifying and resolving power problems late in the flow requires time consuming and expensive iterations. However, designing for low power adds an additional degree of complexity to an already complex problem that has historically targeted performance and area. Optimizing the three variable equation requires a new class of EDA tools that are able to address the power problem from the start to the end of the SoC design flow.

In order to handle complex interrelationships between diverse effects, it is necessary to have a design system whose tools are fully integrated with each other and also with other analysis tools. There are a number of EDA tool vendors that have made, and continue to make, significant contributions in the area of low power design. Two major players in the EDA industry are Cadence and Synopsis. Each has low power flows that provide low power end-to-end solutions from RTL to GDSII. A diverse variety of low power design techniques have been developed to address the various aspects of the power problem.

In addition, comprehensive power reduction also demand a strong ecosystem that includes low power hardware IP blocks, low power libraries, modeling techniques, and silicon wafer fabrication flows that support the low power design methodology.

Two competitive standards for low power SoC design are available. The UPF has been released by Accellera. The CPF is managed by Silicon Integration Initiative's Low Power Coalition. Can two standards survive in this highly competitive market? Possibly yes, however, one will dominate defined by market forces.

Future instantiations of the low power EDA tools will have to address the unique and higher-level power issues associated with analog functionality, packaging, and systems-level challenges.

References

[1] M. Arora. GALS technique fells massive clock trees. *Chip Design Magazine*, http://www.chipdesignmag.com, October/November 2006.

[2] M. Krstic. Request-driven GALS technique for datapath architectures, http://www.deposit.ddb.de/cgi-bin/dokserv?idn=978773063&dok_var=dl&dok_ext=pdf&. filename=978773036.pdf, pp. 12–18, February 2006.

[3] A. Bink, M. de Clercq, and R. York. ARM996HS Processor, ARM White Paper, February 2006.

[4] Cadence Low-Power Methodology Kit. Datasheet, http://www.cadence.com, 2007.

[5] Synopsis Low-Power Solution. White Paper, http://www.synopsis.com, June 2007.

[6] D. Peterman. Unifying the Industry behind a Power Format, http://www.TI.com, October 2006.

[7] Using a Low-Power Kit to Improve Productivity, Reduce Risk, and Increase Quality. White Paper, http://www.cadence.com, 2007.

[8] S. Bailey. Low Power Design Specification from RTL through GDSII, http://www.edadeignline.com, July 2007.

[9] D. Brophy. Accellera Update and Roadmap, http://www.accellera.org, April 2007.

[10] R. Goering. Dueling Power Standard Camps Stall Specification Merger, http://www.eetimes.com, March 2007.

[11] A. Hemani, T. Meincke, S .Kumar, A. Postula, T. Olson, P. Nilsson, J. Öberg, P. Ellervee and D. Lindqvist. Lowering power consumption in clock by using globally asynchronous locally synchronous design style. *Proceedings of AC/WIEEE Design Automation Conference*, 1999.

[12] E. Talpes and D. Marculescu. Toward a multiple clock. Voltage island design style for power-aware processors, *IEEE Transactions on Very Large Scale Integrations (VLSI) Systems*, Vol. 13, No. 5, pp. 591–603, 2005.

CHAPTER 4

Energy Optimized Software

4.1 Mobile Software Platform

The latest networks, radio and hardware technologies that are enabling a multitude of additional services and applications have been considered. One major piece of the puzzle to be explored is the software platform on which modem software and the applications execute.

Software Platforms: The more functionality a mobile device is intended to have, the more likely it is that the device manufacturer has to license, port, and customize a full software platform into the device. The choice of available platforms ranges from operating systems derived from PCs, systems originally designed for handheld organizers, to platforms specifically designed for mobile wireless devices.

Complete mobile device platform architecture must cover network drivers, peripherals, local connectivity, an intrinsically designed voice communication interface, security, browsing, messaging, user interface (UI), packaged applications, and development tools.

A complete platform usually provides choices for operators and developers on how to create applications. Whether to use C/C++, scripting, or Java is a decision to be made according to the application requirements.

Software Components: Smartphones and PDAs rely on operating systems that can be licensed from software vendors, but mobile phones are still largely being built on proprietary operating systems. The internal architectures of phones vary, but they are tuned for voice communications with data communications as another essential function.

The introduction of WAP browsers started the data access era for mobile devices. Browsers also started the trend of licensable software components and porting them to proprietary phone operating systems. This trend will continue as other technologies are made available as licensable components.

The desired functionality of a phone will dictate whether it features MMS, XHTML, WAP, download, DRM, and other security technologies. Java runtime environment allows applications to be downloaded to phones running proprietary operating systems where the applications are executed in a standardized environment.

No matter which software components make up a phone, application developers and service providers benefit from tested and proven pieces of software running in mobile phones, that ensure application software compatibility.

Four principle components of software in a high end cellular handset (Figure 4.1):

1. Modem software that comprises protocol stack, device drivers for platform integrated circuits (ICs), audio, global positioning system (GPS), connectivity, etc.

2. Operating system or kernel.

3. Application framework (middleware), window manager/widgets, gaming engine, synch engines, software configuration database, and some reference applications, personal information management (PIM) applications like dialer and browser.

4. User interface.

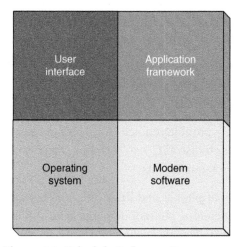

Figure 4.1: Principle Software Components

The software landscape today for mobile devices is more complex. To help to navigate the complex labyrinth of software, Figure 4.2 represents a conceptual model for the mobile phone software stack that makes up a typical handset. Phone software can be visualized as a software stack of functional layers from bottom to top.

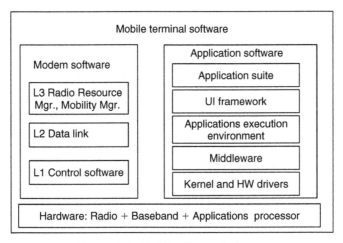

Figure 4.2: Mobile Phone Software

4.1.1 Modem Software

The protocol stack for a Global System for Mobile Communication (GSM) cellular handset follows the basic concepts of the ISO OSI 7 layer stack in so much as it is a layered architecture. The relationship between the layers as defined in the OSI model and those in GSM are described in the following section along with a breakdown of the functions performed by each layer (Figure 4.3).

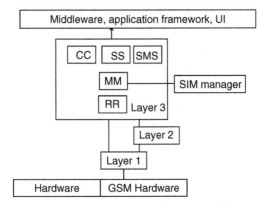

Figure 4.3: The GSM Protocol Stack Architecture

4.1.1.1 Layer 1

In the OSI model, the lowest layer is the physical layer which represents the physical connection between devices. At the lowest level, the physical layer is the radio frequency

(RF) connection between the handset and the base station, but it also includes the radio and baseband hardware which makes this connection possible.

The GSM Layer 1, also called the physical layer or the control software, controls the radio and baseband hardware. In the OSI model this more closely corresponds to the Medium Access Control (MAC) function which is part of Layer 2. However, sometimes the term Layer 1 also covers the signal processing software responsible for equalization of the channel, coding and decoding, etc. Traditionally, the L2 and L3 of the protocol stack has been implemented on a CPU and the signal processing on a digital signal processor (DSP). However, there are a number of implementations appearing now which use a high performance DSP processor where the Layers 1–3 functions are all implemented on the DSP. An example of this implementation is Freescale's MXC family of single core modems (Figure 4.4).

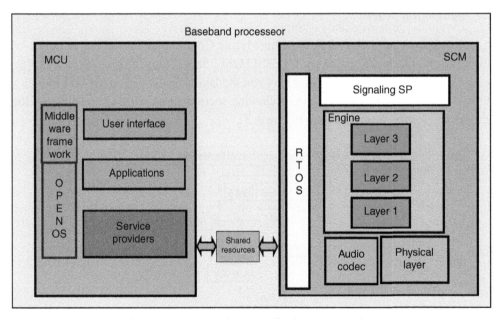

Figure 4.4: Freescale MXC Single Core Modem

Source: http://www.freescale.com

4.1.1.2 Layer 2

Layer 1 multiplexes the physical access to the radio channel and provides a number of logical channels which can be used for signaling. Layer 2 is responsible for establishing a data link on these logical channels to allow reliable transmission of Layer 3 signaling messages. It corresponds to the data link layer (Layer 2) in the OSI model.

Layer 2 can operate in two modes:

1. Unacknowledged mode – used for broadcast data where acknowledgement is not possible and measurement reports where acknowledgement is not essential.

2. Acknowledged mode – used for all connections where reliable transmission of Layer 3 messages is important (e.g. call set-up).

Layer 2 operates by exchanging fixed length frames between two L2 peers (one in the MS and one in the BTS). The structure of the frame is shown in Figure 4.5.

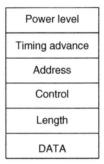

Figure 4.5: Layer 2 Frame

The address field allows for separate connections to Layer 3 for normal signaling and short messages.

The control field distinguishes between different Layer 2 primitives. These enable a link to be established and terminated, allow for error recovery, and contain frame counters that allow for retransmission of eroded frames.

The length specifies the amount of data in the data field. This also allows for a Layer 3 message to be split over a number of Layer 2 frames, although for efficiency reasons, this is not normally required for most messages in the call set-up sequence.

4.1.1.3 Layer 3

Layer 3 is subdivided into a number of separate tasks. These do not follow the OS1 model precisely, but they comprise all of the functions between Layer 2 and the Application layer.

4.1.1.4 Radio Resource Manager

The Radio Resource Manager (RR) is responsible for controlling the aspects relating to the RF connection between the mobile and base station. It is the part of Layer 3 that is most closely coupled to the lower layers.

There is a direct connection between the RR and Layer 1 through which it controls the state of Layer 1 to perform the following functions:

- Channel assignment and release

- Channel change

- Channel mode and ciphering

- Handover between base stations

- Frequency usage, reassignment, and hopping

- Measurement reports

- Power control and timing advance

In addition, the RR is responsible for routing Layer 3 messages received from Layer 2 or Layer 1 to the correct part of Layer 3 by means of the protocol discriminator field in the message.

4.1.1.5 Mobility Manager

The Mobility Manager (MM) is responsible for handling aspects relating to the mobility of the handset:

- Location update when the mobile moves into a new location area and periodically as required.

- Attach/detach when the mobile is powered up and down.

- Authentication in conjunction with the SIM manager.

- Handling the temporary ID (TMSI) in conjunction with the SIM manager.

- Responding to requests for identity (IMSI and IMEI) by the network.

4.1.1.6 Connection Manager

The connection manager (CM) is split into three tasks: call control (CC), supplementary services (SS) and short message services (SMS).

CC is responsible for:

- Call establishment (mobile originated and mobile terminated)
- In call modification of the call type (voice/data)
- Call re-establishment when the call is dropped (if allowed by the network)
- The transmission of DTMF by signaling rather than in-band tones

SS is responsible for providing services such as call barring, call redirection, multi-party calls, etc. in conjunction with the network.

SMS is responsible for the transmission and reception of point-to-point short messages and also for the reception of cell broadcast short messages.

4.1.1.7 Mobile Network

The Mobile Network (MN) provides a non-GSM specific interface between the protocol stack and the application layer. It handles all of the GSM specific functions required by the protocol stack.

4.1.1.8 SIM Interface Manager

The SIM interface manager is responsible for communicating with the SIM card at a procedural level, although the individual transactions are normally handled by a SIM driver. It is activated by a request from either the MM or the Man Machine Interface (MMI) to perform the following functions:

- Perform the authentication algorithm on the SIM using a given random seed (RAND) and return the authentication parameter SRES and ciphering key Kc.
- PIN management
- Storage and retrieval of data, e.g. SMS and phone book

4.1.2 Application Software

The portable device industry has evolved from a view of devices as closed and "dumb" to a view of devices as open platforms that share intelligence with the network. This is part of a general trend toward "openness," which extends to software platforms.

4.1.2.1 What Defines a Good Application Environment?

There are some mobile devices with closed platforms where no additional software can be added after production. However, the trend is toward "open" platforms where third-party software can be downloaded and installed.

Although the concept of openness is well accepted, the wireless mobile industry has not aligned on a standard set of platforms and tools. Operating systems, application execution environments (AEEs) like Java 2 MicroEdition (J2ME), and Binary Runtime Environment for Wireless (BREW), plus value-added platforms like UIQ and Series 60 are just a few examples of the various approaches being taken.

An open operating system can either be a foundation for an applications environment or it can be the applications environment itself. Examples of application environments are the operating systems Symbian, Windows CE, and Linux. We can also include Java and BREW, both of which are independent from the OS. In addition, value-added platforms include the combination of an OS, Java, and MMI.

The high level components of the application environment are described below:

- *Kernel*: The core software which includes hardware drivers, memory, file system, and process management.

- *Middleware layer*: The set of peripheral software libraries which enable handset applications but are not visible to the user. Examples are messaging and communications engines, WAP/page renderers, multimedia codecs, security subsystems, and device management.

- *AEE*: An application manager and set of application programming interfaces (APIs) allowing developers or manufacturers to develop handset applications.

- *UI framework*: A set of graphics components (screens, buttons, lists, etc.) and an interaction framework that gives handset applications their look and feel.

- *Application suite*: The set of core handset applications such as the idle-screen, dialer application launcher or menu screen, contacts, calendar, inbox, browser shell, and settings screens that form the interface that the phone user experiences most of the time.

Figure 4.6 shows the simplified mobile handset software stack, with examples of vendors and products that deliver functionality corresponding to each layer.

However, hundreds of products exist which contribute to the various parts of the software stack. The functionality of each product often spans a number of the stack layers which make understanding software products a complex endeavor.

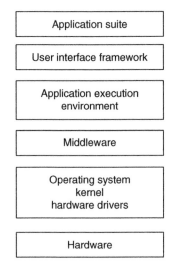

Figure 4.6: Mobile Handset Software Stack

4.1.3 Operating Systems for Mobile Devices

Operating systems are relatively rich software environments. They include UIs, file systems for saving and managing downloaded applications and data, and APIs which provide software developers with re-usable bits of code to handle basic tasks such as drawing on the screen, keyboard input, and telephony functions.

An operating system can be thought of as a "full service" software package. Numerous optional modules can be layered a top operating systems for additional functionality. These can include security and authentication modules and basic applications such as a micro browser or address book. With all the optional modules, the phone can become a very sophisticated device with a wide range of capabilities.

The three biggest mobile operating systems are Windows CE/Pocket PC, Symbian, and Linux OS.

4.1.3.1 Windows Mobile

Windows CE was designed by Microsoft to be a mobile version of its popular desktop OS. Although Windows CE is not a scaled-down version of Windows desktop OS's, the operating systems still have many similarities. Windows CE was indeed developed from scratch in order to fit mobile devices while still keeping most of the look and feel of its bigger desktop brothers. The result was an OS that did not have all of the features of a desktop OS and that still lacked the robustness and speed that was needed for a mobile OS.

Although Windows CE has had a hard time in the consumer segment, vertical appliances such as TV set-top boxes have been really successful. Three versions of Windows CE exist: one for vertical applications such as cars, one for hand-held PCs, and the Pocket PC version.

Although the rich multimedia support and powerful hardware are important strengths of Windows CE, the strongest part is its development environment. Windows CE uses Win 32 APIs and anyone who has been developing for Microsoft's other platforms feel at home with Windows CE. Some Windows programs can actually be recompiled for Windows CE. As always with mobile devices and operating systems, the display and memory properties are significantly different.

The use of multithreading/multitasking significantly increases the possibilities for developers to create good applications. Communication with networked devices can be located on separated threads in order to isolate the rest of the application from disturbances, such as interruptions. One thread can be used to stream stock quotes from a server while another one updates a graph on screen and lets the user manipulate it.

Most Pocket PCs have shorter battery life than the competition. The APIs do support power management functionality and the color screens and powerful CPUs drain power from the battery. Many Windows CE devices have a compact flash or PCMCIA slot that you can also use for external peripherals, such as wireless modems and GPS receivers.

The communications features of Windows CE are intuitive for those who are familiar with the Win32 communication APIs. Most of those APIs have been ported directly to the smaller platform, however, and it is likely that the same code that works for the desktop operating systems will work on a Pocket PC without modifications. This feature gives developers access to a huge library of written code, either through old Win32 projects in house or on the Web. The drawbacks with Windows CE have so far included high battery consumption. This can pose a severe limitation of key metrics such as "talk time" and "PDA" mode.

The latest version, Windows Mobile provides a complete software stack that spans the kernel and middleware to the AEE (supporting both native C++ and the .NET Compact Framework) and the application suite. The .NET Compact Framework is an interpreted environment that simplifies application development.

4.1.3.2 Symbian

From the outset Symbian has been designed as a communications-centric operating system for mobile devices.

Much like Windows CE, Symbian supports multimedia, multitasking and often runs on powerful processors.

Because Symbian was designed to be a good operating system for communicators, there is a dedicated software development kit (SDK), which enables convenient development in many languages. Symbian was the first OS with a WAP browser built in and many developers who start working with Symbian do it for the communication parts and for the support of the wireless giants.

Symbian OS is a semi-complete operating system, providing hardware drivers, kernel, and key middleware components. The OS is offered with complete source code and is (with few exceptions) modifiable by the manufacturer.

The Symbian OS kernel is believed to be one of the most advanced kernels for mobile handsets. Above the kernel sit a number of libraries to support graphics, memory, file systems, telephony, networking, and application management. The middleware layer of Symbian OS includes engines and protocol support for applications, browser, messaging, base UI libraries, codec support and Java support.

The separation of the UI layer dates back to 2001 and was made in the interest of greater manufacturer differentiation. Today, the UI framework, several middleware components, AEE and application suites are provided by S60, UIQ, or MOAP.

4.1.3.3 S60 and Symbian

S60 contains a UI layer, application suite, and middleware engines that run on top of Symbian OS. It is a software platform containing all stack components above the core OS value line. S60's UI framework can be superficially customized by operators and manufacturers, and its application suite which includes voice and video telephony, messaging, contacts, calendar, media gallery, RealPlayer, music player, camcorder, and browser applications. S60 also contains a number of AEEs, programmable via C++, J2ME, FlashLite, and Python. Finally, S60 includes key middleware engines such as SMS, MMS, e-mail transports, SyncML, advanced device management including FOTA, DRM, Bluetooth, and Wi-Fi support.

4.1.3.4 Linux

Linux has long been popular in the IT world for its open source approach, but the widespread adoption of the OS has yet to be seen. With the tightening competition in the mobile Internet device space, the thought of an OS that you can use for free is very appealing. Mobile devices also generally have a lower manufacturing cost than a full-size PC, which means that cutting costs for individual parts is more important.

For mobile phones there are numerous critical Linux components which are not available through the open developer community. By far the most important component is the telephony software responsible for call management and data communication. In addition Linux distributions have not been optimized for mobile devices to address issues such as power management, real-time performance, and start up time.

Also a handset applications suite has to be developed that includes PIM applications, media management, and phone settings. Middleware, such as device management and firmware over the air, Java, and graphics engines must also be included in the Linux distribution.

There is no one single UI for Linux due to its architectural openness.

Linux's open source, decentralized and vendor-independent operational model offers a number of benefits for Linux-based phone product development:

- The Linux kernel has been ported to more than a dozen chipsets, making it the most portable operating system to date.

- Linux is considered low cost and open to innovation.

- Widespread use of embedded Linux in networked appliances makes Linux-based distributions both technically competent and cost-effective for converged devices.

4.1.4 Why an Operating System? Application Execution Environment

This question is becoming highly relevant as we see the migration to a world of standards and platform independence. No one wants to develop an application that can run only on GPRS and Windows Mobile. Developers will design the application to run on all packet-based mobile Internet networks. You can access WAP and other XML applications on Windows devices, and people who use Symbian can also have this access.

Does the applications developer care what the underlying operating system is? From the software developer's point of view, it would be best if the application could be written for one device and one network. The developer would only need to perform minor tweaking in order to make it work on other networks and software platforms.

While this goal might remain a dream, many efforts are being made to consolidate the application environments and to standardize as much as possible. The two leading contenders are the Sun-sponsored Java 2 Micro Edition with Mobile Information Device Profile (J2ME MIDP), and Qualcomm's BREW.

4.1.4.1 Java Implemented in Portable Products

In the early 1990s Java was created to solve the needs of Sun's Green Project. The goal of the Green Project was to develop a set of networked, digitally controlled consumer devices. The Green team could not find a suitable programming language that worked in the diverse set of hardware and programming firmware of the time. Thus, Java was born not out of a need to run independently on thousands of server and client computers as it does today. Instead, it had humble beginnings as a means to network home entertainment equipment.

Recognizing that one size does not fit all, Sun has grouped Java technologies into three editions or formats: the Version 2 Standard Edition (J2SE), an Enterprise Edition (J2EE), and the new J2ME. Each edition is a developer treasure chest of tools and supplies that can be used with a particular product. The J2EE defines the standard for developing multi-tier enterprise applications. The J2SE is the definitive development and deployment environment for enterprise software applications that can run on a variety of desktop computers and servers. The J2ME specifically addresses the vast consumer space, which covers the range of extremely tiny commodities such as smart cards, PDA's, or a cell phone all the way up to a set-top box. There are over a dozen cell phone manufactures alone. Each has different characteristics and interfaces. As you might imagine, the portability of a Java application across such a deep and diverse set of products attracts many to the promises of J2ME. Of course, the challenge for those in the J2ME world is to compact enough of Java's essentials into a very small package.

The J2ME platform is targeted at the two product groups shown in Figure 4.7. The first group contains personal, mobile, connected, information devices, such as cellular phones, pagers, and organizers. The second group includes shared, fixed, connected, information devices, such as set-top boxes, Internet TVs, and car entertainment and navigation systems.

The work on J2ME started with the development of a new, small-footprint Java virtual machine (JVM) called the kilobyte virtual machine (KVM). Two Java Community Process (JCP) standardization efforts, Connected Limited Device Configuration (CLDC) and Mobile Information Device Profile (MIDP), were then carried out to standardize the Java libraries and the associated Java language and virtual machine features across a wide variety of consumer devices. A number of companies participated in these standardization efforts directly, and many more companies and individuals participated indirectly by sending feedback while the standardization efforts were in progress. Major consumer device companies such as Motorola, Nokia, NTT DoCoMo, Palm and Research In Motion (RIM) played a key role in these efforts.

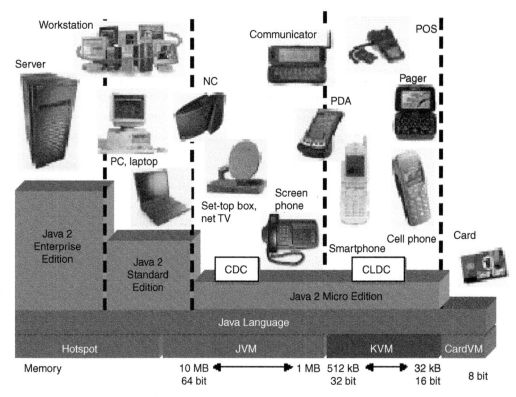

Figure 4.7: Platforms and Target markets

Source: http://www.sun.com

Regardless of the Java environment's format, a compiled Java program, in byte codes, is distributed as a set of class files and is generally run through an interpreter, the JVM, on the client. The JVM converts the application's byte codes into machine level code appropriate for the hardware. The JVM also handles platform specific calls that relate to the file system, the graphical UI, networking calls, memory management that includes garbage collection, exception handling, dynamic linking, and class loading.

4.1.4.2 Binary Runtime Environment for Wireless

BREW includes similar software elements to J2ME CLDC/MIDP and quite a few additional elements as well. Like J2ME, BREW is based on an open programming language, in this case C++. The programming language is well supported by software developers.

The major difference with BREW is that it is more inclusive, including an application testing process and associated software tools for discovering, downloading, selling, and maintaining applications, and a revenue-sharing model. Qualcomm argues that the provision of these features serves the marketplace by eliminating what would otherwise be a lengthy period of trial and error and incompatibility between networks.

4.2 Energy Efficient Software

As described earlier, mobile device software is structured in layers as shown in Figure 4.8. At the bottom there are hardware components such as processors, radios, and power management integrated circuit (PMIC).

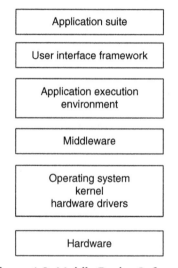

Figure 4.8: Mobile Device Software

OS provides services such as file systems, process control, and memory management. User programs access hardware by issuing system calls through OS. An OS manages resources, including CPU time and memory allocation.

When a user starts a program, a process is created. This process occupies memory and takes CPU time. A process is an instantiation of a program. The life of a process is created, runs, and finally terminates. Most operating systems support multiprogramming: many processes can execute concurrently and share resources. Two processes are concurrent if one starts before the other terminates; namely, their execution times overlap. When a process is alive operating systems manage when it occupies a processor, how much memory it possesses, which files it opens, and which input/output (I/O) devices it uses. Through the services from OS, each process has the illusion that the whole machine is dedicated to it. When multiple processes require the same resource, such as CPU, the operating systems determine their access order; this is called scheduling. Commonly adopted scheduling schemes include round-robin, priority, and first-in–first-out.

4.2.1 Dynamic Power Management

Dynamically determining power states according to workloads is called dynamic power management (DPM). DPM reduces power consumption by dynamically adjusting performance levels to the workload. A general system can be seen as a collection of interacting resources. If workload is non-stationary, some (or many) resources may be idle when the environment does not maximally load the system. The basic idea in DPM is to put underutilized or idle resources into states of operation with reduced or null performance levels that require little or no power.

Most portable devices do not operate at their peak performance continuously. Some devices are idle even when other devices are busy. Occasional idleness provides opportunities for power reduction [1]. DPM puts idle devices into sleeping states to reduce their power consumption [2]. DPM provides the illusion that devices are always ready to serve requests even though they occasionally sleep and save power. DPM can be controlled by hardware, software, or the collaboration of both.

The power manager (PM) is the system component (hardware or software) that performs DPM. A power manager consists of an observer that monitors the load of the system and its resources, a controller, that issues commands for forcing state transitions in system resources, and a policy, which is a control algorithm for deciding when and how to force state transitions (based on information provided by the observer). Power managers determine power state transitions according to their shutdown rules also called policies. Power management includes dynamic voltage scaling and variable frequencies (clock speeds). Setting voltages and frequencies is equivalent to selecting power states.

The two main issues in implementing DPM are: how to design a system that can be efficiently managed with minimal power waste and reduced performance penalty, and how to implement the power manager. Regarding PM implementation, there are three main challenges: controller design, observer design, and policy optimization.

DPM is an independent module of the operating system concerned with active power management. DPM policy managers and applications interact with this module using a simple application program interface (API), either from the application level or the kernel level.

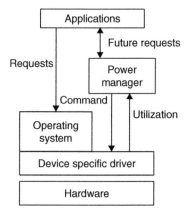

Figure 4.9: DPM Architecture

Although not as broad as the Advanced Configuration and Power Interface (ACPI, discussed later), the DPM architecture does extend to devices and device drivers in a way that is appropriate for SoC's. A key difference with ACPI is the extensible nature of the number of power manageable states possible with DPM (Figure 4.9).

DPM changes the power states of a device based on the variations of workloads. A workload consists of the requests generated from all processes. Figure 4.10 illustrates the concept of power management.

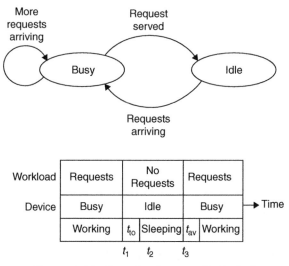

Figure 4.10: Sleep During an Idle Period

When there are requests to serve, the device is busy; otherwise, the device is idle. In this figure, the device is idle between t_1 and t_3. When the device is idle, it can enter a low-power sleeping state.

Changing power states takes time and includes the shutdown and wakeup delays. These delays can be substantial: waking up a display takes several seconds, or hundreds of millions of processor cycles. Furthermore, waking up a sleeping device may take extra energy. In other words, power management has overhead. If there were no overhead, power management would be simple as shutting down a device whenever it is idle. Unfortunately, there is overhead; a device should sleep only if the saved energy can justify the overhead. The rules to decide when to shut down a device are called power management policies.

The break-even time (T_{be}) is the minimum length of an idle period to save power. It depends on the device and is independent of requests or policies.

Consider a device whose state-transition delay is T_o and the transition energy is E_o. Suppose its power in the working and sleeping states is P_w and P_s, respectively. Figure 4.11 shows two cases: keeping the device in the working state or shutting it down. The T_{be} makes energy in both cases equal; it can be found by $P_w \times T_{be} = E_o + P_s \times (T_{be} - T_o)$. Also, the T_{be} has to be larger than the transition delay; therefore,

$$T_{be} = \max(E_o - P_s \times T_o/P_w - P_s, T_o)$$

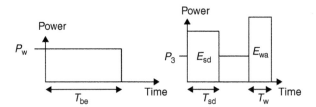

Figure 4.11: Device in Working State Versus Shutting the Device Down

DPM algorithms, also known as policies, can be divided into two major categories: predictive and stochastic.

Predictive policies explicitly predict the length of an idle period before it starts. If the predicted length is larger than T_{be}, the device is shut down immediately after it becomes idle. These policies compute the length of an idle period according to previous idle and busy periods [3–5]. The major problem is that some policies have low prediction accuracy.

Stochastic policies use stochastic models for request generation and device state changes. Under such formulation, power management is a stochastic optimization problem. These policies can explicitly trade off between power saving and performance while predictive policies cannot. Stochastic policies include discrete-time and continuous-time stationary models [6, 7], time-indexed semi-Markov models [8], Petri Nets [9], and non-stationary models [10]. A common problem is that most policies require the characteristics of the workloads for off-line optimization. There is no universally adopted method to adjust at run time if the workload behavior changes.

4.2.2 Energy Efficient Compilers

Software does not consume energy. However, the execution and storage of software requires energy consumption by the hardware. Software execution corresponds to performing operations on hardware, as well as accessing and storing data. Thus, software execution involves power dissipation for computation, storage, and communication.

Storage of computer programs in semiconductor memories requires energy. Dynamic power is consumed to refresh DRAMs and static power is consumed for SRAMs. The energy budget for storing programs is typically small and predictable at design time. Reducing the size of programs, the usual goal in compilation, correlates with reducing their energy storage costs. Additional reduction of code size can be achieved by means of compression techniques.

Energy efficient compilers reduce memory accesses and consequently reduce energy consumed by these accesses. One way to reduce memory accesses is to eliminate redundant load and store operations. Another is to keep data as close as possible to the processor, preferably in the registers and lower-level caches, using aggressive register allocation techniques and optimizations improving cache usage. In addition several loop transformations alter data traversal patterns to make better use of the cache. Examples include loop interchange, loop tiling, and loop fusion. The assignment of data to memory locations also influences how long data remains in the cache.

The energy cost of executing a program depends significantly on its machine or compiled code. For any given architecture, energy cost is tied to machine code. Since the machine code is derived from the source code from compilation, it is the compilation process itself that affects energy consumption.

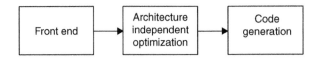

Figure 4.12: Major Tasks in a Software Compiler

The energy cost of machine code can be affected by the back-end of software compilation (Figure 4.12). In particular, energy is affected by the type, number, and order of operations and by the means of storing data, e.g., locality (registers versus memory arrays), addressing, and order.

4.2.2.1 Software Compilation

Most software compilers consist of three layers: the front-end, the machine-independent optimization, and the back-end. The front-end is responsible for parsing and performing syntax and semantic analysis, as well as for generating an intermediate form, which is the object of many machine-independent optimizations. The back-end is specific to the hardware architecture, and it is often called the code generator.

Energy efficient compilation is done by introducing specific transformations in the back-end as they are directly related to the underlying architecture. Some machine-independent optimizations can be useful in general to reduce energy consumption. An example is selective loop unrolling, which reduces the loop overhead, but is effective if the loop is short enough. Secondly, software pipelining decreases the number of stalls by fetching instructions from different iterations. Finally, removing "tail recursion" eliminates the stack overhead.

The main tasks of a code generator are instruction selection, register allocation, and scheduling. Instruction selection is the task of choosing instructions, each performing a fragment of the computation. Register allocation is allocating data to registers. Instruction scheduling is ordering instructions in a linear sequence.

When considering compilation processors, instruction selection and register allocation are often achieved by dynamic programming algorithms [11], which also generate the order of instructions. Instruction selection, register allocation, and scheduling are intertwined problems that are much harder to solve [12].

The traditional compiler goal is to speed up the execution of the generated code by reducing code size. Executing machine code of minimum size consumes minimum energy if we neglect the interaction with memory and assume a uniform energy cost for each instruction.

Energy efficient compilation strives to achieve machine code that requires less energy, compared to a performance-driven traditional compiler, by leveraging the non-uniformity in instruction energy costs and the energy costs for storage in registers and in main memory, due to addressing and address decoding. Nevertheless, results are sometimes contradictory.

For some architecture, energy efficient compilation gives a competitive advantage compared to traditional compilation, for some others the most compact code is also the most economical in terms of energy, thus obviating the need for specific low-power compilers.

Register assignment aims at the best utilization of available registers by reducing spills to main memory. Registers can be labeled during the compilation phase, and register assignment can be done with the goal of reducing switching in the instruction register as well as in register decoders [13].

When considering embedded system applications, the memory/processor bandwidth can be reduced by recoding the relevant instructions and compressing the corresponding object code.

4.2.2.3 Static Compilers

Compiler-based approaches are hindered by the compiler's view that is limited to the programs it is compiling. This raises two problems. First challenge is that compilers have incomplete information, such as the control flow paths and loop iterations that will be executed, regarding how a program will actually behave. Statically optimizing compilers often rely on profiling data that is collected from a program prior to execution, to determine what optimizations should be applied in different program regions. A program's actual runtime behavior may diverge from its behavior in these "simulation runs," thus leading to suboptimal decisions.

The second challenge is that compiler optimizations tend to treat programs as if they exist in a vacuum. While this may be ideal for specialized embedded systems where the set of programs that will execute is determined ahead of time and where execution behavior is mostly predictable, real systems tend to be more complex. The events occurring within them continuously change and compete for processor and memory resources.

4.2.2.4 Dynamic Compilers

Dynamic compilation addresses some of these problems by introducing a feedback loop. A program is compiled but is then also monitored as it executes. As the behavior of the program changes, the program is recompiled to adapt to these changes. Since the program is continuously recompiled in response to runtime feedback, its code is of a significantly higher quality than it would have been if it had been generated by a static compiler.

There are a number of scenarios where continuous compilation can be effective. For example, as battery capacity decreases, a continuous compiler can apply more aggressive transformations that trade the quality of service for reduced power consumption. However, the cost for a dynamic compiler needs to perform a cost benefit analysis to insure that the cost of recompiling a program is less than the energy that can be saved.

4.2.2.5 Other Compiler Techniques

Additional compiler optimizations, which are done to minimize energy consumption, are loop unrolling and software pipelining to increase instruction level parallelism.

Loop Unrolling

Loop unrolling is one of the techniques to increase instruction level parallelism by decreasing the number of control statements which execute in one loop by creating longer sequences of straight-line code. Loop unrolling is controlled by the unrolling factor, which is the number of duplications of the body statements inside the loop.

Loop unrolling decreases the number control operations and overhead instructions. The overhead is code space. There is a diminishing return, however, from excess unrolling.

As the loop unrolling factor increases, the average switching energy per cycle increases while the total energy of the program decreases.

Software Pipelining

Software pipelining is a technique for re-organizing loops such that each iteration in the software-pipelined loop is formed from instruction sequences chosen from different iterations in the original code segment. The major advantage of software pipelining over straight loop unrolling is that software pipelining consumes less code space.

Software pipelining decreases the number of stalls by fetching instructions from different iterations. Hence, total energy consumption reduces due to this reduction in stalls and furthermore the program takes fewer cycles to finish.

Eliminating Recursion

Compilers usually execute recursive procedures by using a stack that contains pertinent information, including the parameter values, for each recursive call. Good compilers usually provide a feature called tail recursion in which the recursion occurring at the end can be eliminated. Recursion elimination saves much of the overhead usually associated with calls. This is also true for inline expansion of a subprogram. Other optimizations become possible because the actual parameters used in a call become visible in the subprogram body. Especially in the case of actual parameters that are literals, opportunities for folding and deleting of unreachable code can be expected. The total switched capacitance drops significantly as recursion decreases, however, the switched capacitance per cycle (power) increases because the number of expensive memory and register overhead operations decreases and hence the number of cycles which use the stack decreases.

4.2.3 Application-Driven Power Management

Why not let the application programs control the service levels and energy cost of the underlying hardware components? There are two objections to such an approach. First, application software should be independent of the hardware platform for portability reasons. Second, system software supports generally multiple tasks. When a task controls the hardware, unfair resource utilization and deadlocks may become serious problems.

For these reasons application programs contain system calls that request the system software to control a hardware component, e.g., by turning it on or shutting it down, or by requesting a specific frequency and/or voltage setting. The request can be accepted or denied by the operating system that has access to the task schedule information and to the operating levels of the components. The advantage of this approach is that OS-based power management is enhanced by receiving detailed service request information from applications, and thus is in a position to take better decisions.

Prior to application-driven power management there were two standard approaches defined to reduce power management, Advanced Power Management (APM) and Advanced Configuration and Power Interface (ACPI).

4.2.4 Advanced Power Management

APM was the first standard designed to reduce power consumption in personal computers. It defines four power states:

1. Enabled
2. Standby
3. Suspend
4. Off

APM uses activity timeouts to determine when peripherals should be switched to a lower state. Though still found in most PCs, APM has several shortcomings. For one thing, it runs in the BIOS, a component rarely used in embedded systems. It also makes power management decisions without informing the OS or user applications. In fact, it will sometimes misinterpret user actions and place the system into the Suspend state, even though the user is still interacting with the system. In addition APM is very specific to each system and as a result it is very difficult to port and extend.

4.2.5 Advanced Configuration and Power Interface

A new specification for power management, called the ACPI, was introduced in 1997 by Compaq, Intel, Microsoft, and Toshiba. ACPI provides a uniform hardware/software (HW/SW) interface for power management and allows hardware vendors, operating system (OS) designers, and device driver programmers to use the same interface. A hardware device complies with ACPI if it can properly respond to ACPI calls such as setting and querying power states.

Designed to overcome the drawbacks of APM, the ACPI specification can be divided into two parts: an implementation at the hardware and BIOS level, and functionality incorporated into the OS itself. In a nutshell, tables in the BIOS define the power modes for individual peripherals, and the OS uses these definitions to decide when to switch a device, or the entire system, from one power state to another.

Figure 4.13 shows the ACPI interface [14] in which the power manager resides in the operating system. ACPI controls the power states of a whole system as well as the power states of each device.

ACPI defines the following:

- Hardware registers – implemented in chipset silicon
- BIOS interfaces
    - Configuration tables
    - Interpreted executable function interface (Control Methods)
    - Motherboard device enumeration and configuration
- System and device power states
- ACPI Thermal Model

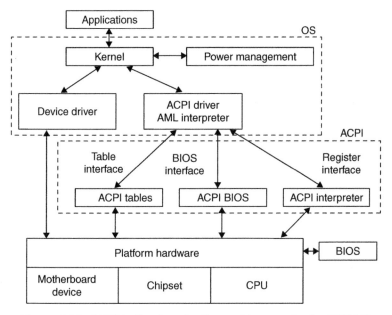

Figure 4.13: ACPI Indicating the Power Manager in the OS [14]

ACPI moves power management decisions out of the BIOS and into the OS, allowing a system to support power management policies that are both more sophisticated and more effective. Giving this responsibility to the OS is an approach designed for systems that are typically "on." It does not apply to something like a portable device, which requires intelligent power conservation when the power switch is "off." As discussed, such a system requires a power policy that can respond to battery level, passage of time, and other application-specific scenarios. An OS, by nature, has far too little knowledge of such scenarios to make the appropriate power management decisions.

4.2.6 The Demand for Application-Driven Power Management

Approaches that rely on the BIOS or OS cannot accommodate the many unique requirements of cellular handsets, portable medical instruments, networking elements, tactical field radios, and other power-sensitive devices. The need is for application-driven power management which includes:

- The power policy is directed by a separate, user-space process, not the OS, giving more control to the application developer.

- The power policy can define a superset of the power states supported by standards like ACPI, providing fine-grained control over the power level of the entire system and of individual peripherals.

- The power policy can account for external events unanticipated by existing standards, including events whose semantics are known to the application developer but invisible to the OS or BIOS.

To achieve this level of control, developers need to build a framework that provides a flexible set of mechanisms for managing power. Figure 4.14 shows how such a framework has been implemented.

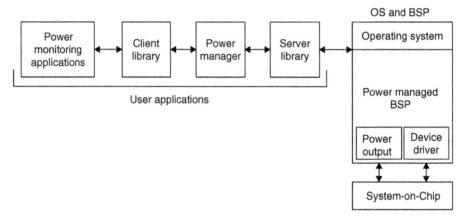

Figure 4.14: Application-Driven Power Management Implemented in a Client–Server Architecture in Which a Power Manager Interacts with Applications and Peripherals

4.2.6.1 Power Manager

The power manager is a user-space process that implements a power policy defined by the system designer. This policy is aware of the power states (Active, Idle, etc.) supported by each power managed entity like peripherals and power-sensitive applications. The power manager implements the rules that determine exactly when each entity will transition from one power state to another.

The power manager is a "power broker" that uses its knowledge of all subsystems and applications to negotiate the most appropriate power-state changes for any given scenario. For example, if an application no longer needs a peripheral, the power manager could check whether other applications need that peripheral before deciding to power it down.

4.2.6.2 Power Monitoring Applications

These user-written applications monitor any aspect of the system and notify the power manager of data or events that may require a change in the system's power state. Using this information, the power manager can re-evaluate the power state and modify it.

4.2.6.3 Power-Sensitive Applications

Programs, both applications and system services, modify their behavior according to the system's power state. If the state changes, these applications will receive notification from the power manager and enable or disable features consistent with the new state. Conversely, an application could ask the power manager to change the power states of other entities.

4.2.6.4 Drivers for Power Managed Peripherals

When the power manager determines that a peripheral's power state must change, it notifies the appropriate device driver, along with any other interested clients. The driver can then change the power state accordingly. To make fine-grained power management decisions, the power manager must know the exact power states supported by each peripheral. It could learn of these states when the device driver starts or during power manager initialization.

4.2.6.5 Client/Server Libraries

These libraries provide the necessary glue that allows the power manager and its clients (e.g. power monitoring applications) to communicate with each other. The server library would also provide building blocks for creating a power manager, including APIs for defining the power policy.

4.2.6.6 Addressing the Challenges

The increased control provided by an application-driven model of power management can address the challenges of predictable response times and the need to respond to external stimuli over a prolonged time. Consider the issue of predictable response time. An approach like ACPI can easily power down a peripheral after a period of inactivity, not realizing that an application needs the peripheral to meet an important, and imminent, deadline. With an application-driven approach, on the other hand, the developer can build checks into the power manager to ensure that this scenario does not occur. The power manager could query every power-sensitive application before deciding whether the peripheral should be powered down.

Maintaining prolonged operational readiness, the difference between an application-driven framework and approaches like APM and ACPI are worthy of consideration. In a portable device these approaches have no concept of the devices involved, such as the modem, or exactly how and when certain peripherals should be powered down to ensure long-term readiness. With an application-driven framework, however, all this information can be built into the custom power policy. Developers can control exactly when to step down the power consumption of any given peripheral, be it minutes, hours, or even weeks after the portable device has been turned off. Table 4.1 compares the different power management approaches.

Table 4.1: Comparison of Power Management Approaches

Technique	Process	Applicability
APM	BIOS control	BIOS rarely employed in portable devices
	Employs activity timeouts to determine when to power down a device	APM makes power-management decisions without informing OS or individual applications
ACPI	Decisions managed through the OS	Has no knowledge of device-specific scenarios. Enables power policies for computing systems with standard usage patterns and hardware
	Control divided between BIOS and OS	Enables sophisticated power policies for general-purpose computers with standard usage patterns and hardware
Application-driven power management	Provides flexible mechanisms, not predefined power policies, to manage power	Enables fine-grained control over power states of individual components
	Moves control out of BIOS/OS and into a customizable user-space process	Addresses external events known to the application and invisible to the OS
		Accommodates unique, device-specific requirements (e.g. predictable response times)
		Can support APM or ACPI if required

4.2.6.7 Putting the Developer in Control

When it comes to power management, no two mobile systems are the same. Even within a single product line, systems may use different power sources, peripherals, and CPUs; they may also support different operational scenarios. Together, these factors result in a host of application-specific requirements that existing power management standards, with their focus on general-purpose computers, have not anticipated.

A framework that allows customizable user-space processes to direct power management is much better equipped to handle these myriad requirements. Using such a framework, system designers can create a power policy for each mobile system they deploy and, as a result, ensure faster response times, a wider range of operational scenarios, and, of course, longer battery life.

4.3 Summary

While operating systems are the core of the Microsoft and Symbian strategies, they tend to reinforce a need for expensive hardware. But they can be stripped down and eventually the increasing capabilities of silicon will make them less expensive. Client-side execution environments, notably J2ME and BREW, provide an alternative. They encourage non-proprietary application development. All leading phone operating systems will support client-side execution, but an operating system is not required to implement client-side execution. However, industry polarization on J2ME or BREW may prove to be a problem for widespread interoperability.

A number of techniques are available to the developer of mobile devices to insure that energy efficient software is a key driver. These techniques range from energy efficient compilation, use of standard specifications like APM and ACPI, and more sophisticated techniques like DPM and its extensions.

The Increasing Value of Software: Code size growth is exploding in the mobile devices. Figure 4.15 indicates how code size has evolved from 2G to 3G cellular products.

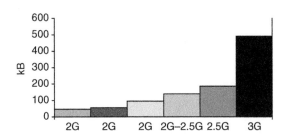

Figure 4.15: Code Size Growth from 2G to 3G

It is clear that many of the complexities, either from a technology or changing standards, drive the need for more functionality realized in software. Figure 4.16 demonstrates the transition the wireless mobile industry is undertaking.

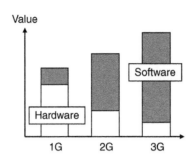

Figure 4.16: Hardware to Software Transition

Software is indicated by the patterned boxes and hardware by the clear boxes in Figure 4.16 for 1G, 2G, and 3G mobile terminals, respectively. The reduction on the importance of hardware is due to the integration efforts ongoing in the semiconductor industry previously discussed. The exponential increase in software is due to the multiple protocols, platform software, applications and middleware required for future wireless mobile products.

References

[1] R. Golding, P. Bosch, and J. Wilkes. Idleness is not sloth. *Proceedings of the USENIX Winter Conference*, New Orleans, LA, January 1995, pp. 201–212.

[2] L. Benini, and G. De Micheli. *Dynamic Power Management: Design Techniques and CAD Tools*. Norwell, MA: Kluwer, 1997.

[3] M.B. Srivastava, A.P. Chandrakasan, and R.W. Brodersen. Predictive system shutdown and other architecture techniques for energy efficient programmable computation. *IEEE Transactions on VLSI System*, Vol. 4, March 1996, 42–55.

[4] E.-Y. Chung, L. Benini, and G. De Micheli. Dynamic power management using adaptive learning tree. *International Conference on Computer-Aided Design*, San Jose, CA, November 1999, pp. 274–279.

[5] C.-H. Hwang, and A.C.H. Wu. A predictive system shutdown method for energy saving of event driven computation. *ACM Transactions Design Automation Electron System*, Vol. 5, No. 2, 2000, 226–241.

[6] L. Benini, A. Bogliolo, G.A. Paleologo, and G. De Micheli. Policy optimization for dynamic power management. *IEEE Transactions on Computer-Aided Design Integrated Circuits System*, Vol. 16, June 1999, 813–833.

[7] Q. Qiu and M. Pedram. Dynamic power management based on continuous-time Markov decision processes. *Proceedings of the 36th ACM/IEEE Conference on Design Automation*, New Orleans, LA, 1999, pp. 555–561.

[8] T. Simunic, L. Benini, P.W. Glynn, and G. De Micheli. Dynamic power management for portable systems. *Proceedings of the International Conference on Mobile Computing Networking*, Boston, MA, August 2000, pp. 11–19.

[9] Q. Qiu, Q. Wu, and M. Pedram. Dynamic power management of complex systems using generalized stochastic Petri nets. *Proceedings of the Design Automation Conference*, Los Angeles, CA, June 2000, pp. 352–356.

[10] E.-Y. Chung, L. Benini, A. Bogliolo, and G. De Micheli. Dynamic power management for nonstationary service requests. *Proceedings of the Design Automation Test Europe*, Munich, Germany, March 1999, pp. 77–81.

[11] A. Aho, R. Sethi, and J. Ullman. *Compilers: Principles, Techniques, and Tools.* Reading, MA: Addison-Wesley, 1986.

[12] G. Goossens, P. Paulin, J. Van Praet, D. Lanneer, W. Guerts, A. Kifli, and C. Liem. Embedded software in real-time signal processing systems: design technologies. *Proceedings of the IEEE*, Vol. 85, No. 3, March 1997, 436–454.

[13] H. Mehta, R.M. Owens, M.J. Irwin, R. Chen, and D. Ghosh. 1997. Techniques for low energy software. In B. Barton, M. Pedram, A. Chandrakasan, and S. Kiaei (Eds), *Proceedings of the 1997 International Symposium on Low Power Electronics and Design*, ISLPED '97, Monterey, CA, August 18–20. ACM Press, New York, pp. 72–75.

[14] L. Benini and G.D. Micheli. *Dynamic Power Management: Design Techniques and CAD Tools.* Boston, MA: Kluwer, 1998.

CHAPTER 5

Batteries and Displays for Mobile Devices

5.1 Introduction

Two of the most important components today in a mobile device are the battery and the display. It can be argued, that if battery and display technology had not evolved as fast as it did these last few years, features like color displays, high-speed data applications, assisted global positioning system (GPS), and 3D graphics would not be available in a portable device.

5.1.1 Battery Challenge

As mobile devices evolved from analog to digital technology, batteries for mobile devices evolved from sealed lead acid (SLA) to nickel metal hydride and subsequently to lithium ion. Since handset power demands were dropping, at the same time that battery energy densities were rapidly increasing, all was well, and both standby times and talk-times shot up like a rocket. Today it's not unusual for digital handsets to have standby time in excess of a week, and customers have come to expect this.

The problem is that batteries are having difficulties in keeping up with today's mobile handset power demands. As handsets acquire color screens, cameras, and broadband access technologies, power demands have been quickly inching up, but battery storage capacities are not keeping pace. For example, the first 3G W-CDMA handsets introduced in Japan by NTT DoCoMo had standby times typically less than 24 h. The performance gap was an issue to the Japanese accustomed to standby times of almost 2 weeks. As a result 3G broadband wireless technologies have been slow to be accepted by the consumer.

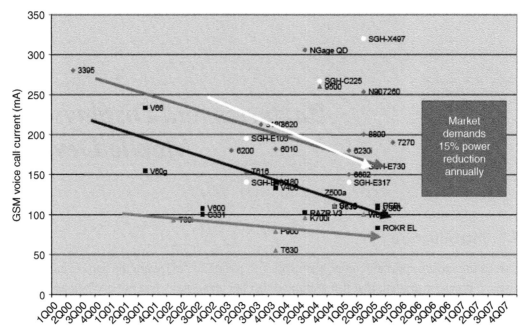

Figure 5.1: Global System for Mobile Communication (GSM) Handset Talk Power

Source: http://www.phonescoop.com

Future battery developments will be evolutionary rather than revolutionary. However, consumers should give up hope of ever seeing long lasting wireless devices in the near future.

Figures 5.1 and 5.2 should be of no surprise to those in the handset battery industry.

"Standby" and "talk" power consumption is decreasing in handsets. However, power consumption during standby or talk does not tell the full story, since many handset feature, such as color liquid–crystal displays (LCDs) (and backlighting), GPS, and wireless data connectivity, are not active during these states. Fortunately for handset makers, battery technology has continued to advance, so battery size or weight does not have to increase along with battery capacity. However, there are some situations where power demands have increased so quickly that battery technology has not been able to keep up, for example multi-mode 3G W-CDMA handsets.

The power consumption of handsets has been steadily decreasing over the years; from the early days of the 10 lb. SLA battery-powered mobile phone. Today's handsets, in both standby and while connected during a conversation, draw only a fraction of the

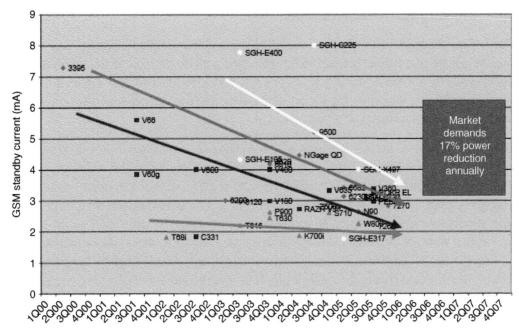

Figure 5.2: Global System for Mobile Communication (GSM) Handset Standby Power

Source: http://www.phonescoop.com

power of their predecessors. For example, the average power consumption of a handset in standby was 257 mW. In 2002 that same average power consumption had dropped to approximately 14 mW, less than 1/18th of the power.

Some lithium battery usage examples are shown below:

- Capacity: 270–300 Wh/l
- Battery sizes: $50 \times 30 \times 5$ mm $= 7,500$ mm^3
- Battery capacity $= 675$ mAh
- Average talk current $= 200$ mA
- Average standby current $= 2$ mA
- Use profile: 1-h talk and 12-h standby $\sim$225 mAh
- Usage between power re-charge $= 3$ days
- Add media playing: 350–450 mAh
- Usage between power re-charge $\sim$1 day

5.1.2 Evolution of Battery Technology

The first handset batteries were SLA, and the battery alone weighed 30 times more than the weight of a typical handset today. But even with that weight, cellular handsets would typically only have standby times of a few hours, and talk-times were measured in minutes. Today, standby time for a handset not weighing much more than a few ounces can be in excess of a week.

From SLA batteries to today's *lithium-ion* (Li-ion) batteries, mobile devices evolved and incorporated several different battery types. First, after lead acid was *nickel cadmium* (Ni-cad). These batteries appeared in mobile devices at the same time that many products were going "rechargeable." This included flashlights, grass clippers, razors, computers, and other portable electronic products. Ni-cad batteries worked well in many applications, but for devices that were not fully charged before recharging, or for uses where batteries sat unused for long periods, Ni-cad batteries would internally short, or sometimes exhibit "memory effects" which would greatly reduce their life. If they were used heavily, they were great. However, in applications like cellular handsets, they were not ideal. In addition, considering the large number of cellular handsets that were disposed of every year, disposing of Ni-cad batteries containing toxic metals was a great concern.

The next battery to appear in a cellular handset was the *nickel metal hydride* (Ni-mh) battery. These batteries were much less toxic than nickel cadmium batteries, in addition to having a greater capacity for their weight, and minimal claimed "memory effect." Ni-mh batteries had a flaw, a high self-discharge rate. If a mobile deice with a nickel metal hydride battery were left in a drawer uncharged for 2 months, even with the device turned off, the battery would likely go dead. If your mobile device was charged everyday, this would not be a problem, but for an occasional use mobile device, the high self-discharge rate of nickel metal hydride batteries was a big problem.

Another solution was needed, and that solution was a rechargeable version of the lithium battery used by watches and cameras, the Li-ion battery. Not only could a Li-ion battery hold a charge for a long time, it also has no memory effect. It is light, and has a very high capacity for its weight, an ideal power source for a portable mobile device. Li-ion batteries, if they are not charged in a precise manor, or are overcharged, can explode or catch on fire. They contain lithium, which is a highly volatile metal that burns when put in contact with water.

Battery manufacturers incorporate several safety features into their batteries to abate the concerns of exploding Li-ion batteries. A typical Li-ion safety circuit must protect a battery against over-current charge, over-voltage charge, over-temperature, and under-voltage discharge.

Lithium-ion polymer batteries appeared next. They are touted as being safer than lithium ion because their electrolyte is composed of a dry polymer. In addition, rather than being built using a conventional metal cell, they use a type of Mylar "bag," making them much more shapeable.

Unfortunately, their charge efficiency is less than conventional lithium ion.

The next energy source for a mobile device may not be a true battery at all, but rather a *fuel cell*. Rather than charging a fuel cell, it's filled with some type of fuel, typically methanol or ethanol alcohol. When the fuel is gone, you just refill it, like the gas tank on your car. But fuel cells, unlike automotive internal combustion engines, do not produce poisonous fumes. Rather they emit water and carbon dioxide, much like human breath. Fuel cells are claimed to run a cellular handset for 3 to 10 times as long as a lithium-ion battery, weighing the same.

5.2 Battery Fundamentals

A battery converts chemical energy into electrical energy. It consists of two electrodes that are electrically connected to active materials, immersed in electrolyte with porous separator placed between them to prevent electric contact, but allow ionic flow (Figure 5.3).

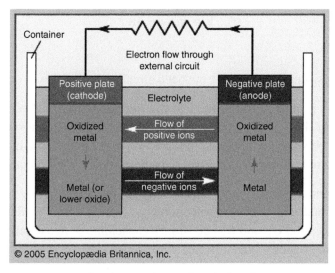

© 2005 Encyclopædia Britannica, Inc.

Figure 5.3: Electrochemical Cell

Source: http://www.britannica.com

The positive (+) electrode, called the cathode, is connected to an aggressive oxidizing material that is capable of ripping electrons from other materials. One example of a common oxidizing material is oxygen which is used in cathodes of fuel cells. However, in batteries solid oxidizing materials, such as MnO_2, and $NiO(OH)_2$, are used.

The negative (−) electrode, called the anode, is connected to a strong reducing material, that is rich in electrons which are easily released. These materials are similar to fuel in its function. Indeed some common fuels like natural gas can be used as an anode agent of fuel cells. In batteries it is more practical to use solid fuels, such as Cd or Zn metals, or more exotic lithium intercalated into graphite.

Batteries have a fixed amount of active materials, which are physically attached to the electrodes. A rechargeable battery retains its crystal structure and electric connecting over many charge and discharge cycles, but eventually degrades.

From an electrical point of view, batteries are often represented either as simply a voltage source, or as voltage source connected in series with a resistor representing internal resistance of the battery. Constant voltage is accurate only for given state of charge of the battery with zero current. When the battery is charged or discharged, its open circuit voltage changes, therefore it can be electrically represented as a capacitor with variable capacitance. When current is flowing in or out of the battery, the battery terminal voltage is described in the following equation:

$$V = Vocv - \text{IR}$$

where

R is the internal resistance and $Vocv$ is the open circuit voltage at given state of charge.

However, a battery is not a capacitor because the electron transfer between materials has to go through many different electrical and chemical stages. Electrical stages include electrons moving between particles and ions moving in the pores. Chemical stages are the reaction on the particle surface and diffusion inside particles of the electrode material.

In Figure 5.4, R_c and R_a are summary diffusion, conduction, and charge transfer resistances for cathode and anode correspondingly. C_c and C_a are chemical charge storage capacitances and C_{c2} and C_{a2} are surface capacitances. R_{ser} is serial resistance that includes electrolyte, current collectors, and wires resistances.

Each stage is associated with its own time constants, which causes complex electrical behavior. To represent battery transient behavior correctly, use an equivalent circuit rather

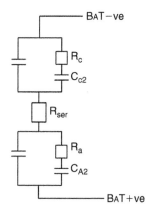

Figure 5.4: Circuit Equivalent of Battery Conduction Under Variable Load Conditions

than simple resistance. The simple circuit in Figure 5.4 is given as an example and is valid for time constants between 1 Hz and 1 MHz. For higher frequencies, additional inductive and capacitive elements need to be added. Different battery types need to use different equivalent circuits if very accurate representation is required.

5.3 Battery Technologies

5.3.1 Sealed Lead Acid

Typically 2 to 3 AH in size, these SLA batteries can power a 3-W transportable cellular for around 12 h of standby time and 40–60 min of "talk-time." Today most of these transportable phones have disappeared, as most users have long switched to a much smaller and lighter digital handset, but some do exist in rural areas where it might be some distance to the nearest cell tower.

5.3.2 Nickel Cadmium

When one thinks of rechargeable batteries for portable devices, probably the first battery to come to mind will be the nickel cadmium battery. This battery has been used in so many different types of rechargeable devices, that many people assume that all rechargeable batteries are nickel cadmium batteries. Although this is not the case, the reliability and robustness of nickel cadmium batteries has to be given much of the credit for the popularity of the rechargeable battery concept.

5.3.3 Nickel Metal Hydride

Nickel Cadmium batteries had been very reliable in many portable applications; however, a cellular phone is a much different application than many applications. For one thing, it is typically placed in its charger long before its full charge is exhausted. As noted in the previous section, this type of charging routing is not the best in terms of reducing the memory effect. Because of this, work was done to find a battery with less memory effect, and perhaps one that was a bit lighter as well. This is where the nickel metal hydride battery comes in.

5.3.4 Lithium Ion

Without question, today's mobile wireless devices rely on one type of battery more than all the others combined, and that battery chemistry is lithium ion. Pioneering work for the lithium battery began back in 1912, but it was not until the early 1970s when the first non-rechargeable lithium batteries became commercially available. Attempts to develop secondary (rechargeable) lithium batteries followed in the 1980s, but failed due to safety problems.

Lithium is the lightest of all metals, has the greatest electrochemical potential, providing the largest potential energy content, but lithium is also an extremely volatile substance that can burn when exposed to moisture.

5.3.5 Lithium-Ion Polymer

A variant of the lithium-ion technology is lithium-ion polymer (Li-poly), which is usually and incorrectly referred to as lithium polymer. The lithium-ion polymer is a natural for portable devices, because it can be produced in a very thin, shapeable package. However, it has not gained the commercial success as predicted. The problem is cost and tough competition from prismatic lithium-ion cells. While lithium-ion polymer batteries can still be made thinner and lighter than prismatic batteries, most lithium-ion batteries are thin enough to meet current handset requirements.

5.3.6 Other Lithium-Ion Types

Lithium-ion cells, both cylindrical and prismatic continue to evolve, and in many cases meet or exceed the advantages of lithium-ion polymer, without the high cost or reduced efficiency. For example, since last year, Toshiba has been manufacturing a battery it calls an advanced lithium battery (ALB). The ALB is a cross between a conventional lithium-ion cell and a lithium-ion polymer cell. These cells can be made as thin as 1 mm while still achieving a volumetric energy density of greater than 365 Wh/l.

5.4 Battery Chemistry Selection

Many new challenges are arising for developers, due to higher current drains and greater temperature extremes. Every portable device has a unique set of power drain and temperature profiles that must be characterized, understood, and addressed during the development process. The usage profile includes temperature ranges, discharge profiles, charging regimens, expected shelf life, and transportation requirements, and should account for foreseeable misuse as well as intended use. Both external and internal operating temperatures are important considerations in selecting the optimal cell for a portable application. Many portable devices are expected to operate in a range from −40 to +60°C. A summary of battery chemistry selection is illustrated in Table 5.1.

Table 5.1: Chemistry Selection [1]

	SLA	Ni-Cd	Ni-mh	Li-ion
Operating voltage (V)	2	1.2	1.2	3.6
Maximum series cells (V)	12 (24 V)	10 (12 V)	10 (12 V)	7 (25 V)
Energy density	35 Wh/kg 70 Wh/l	40 Wh/kg 125 Wh/l	70 Wh/kg 200 Wh/l	175 Wh/kg 500 Wh/l
Cycle life	500–800	500–1,000	500–1,000	500
Advantage	Cost	High drain rates	Good value	High energy density
Disadvantage	Low capacity	Low capacity	Venting	Complicated packs

High currents and temperature extremes also harm the battery system's electronics, damaging key components. Temperature extremes can cause similarly rated battery cells from different vendors, to demonstrate widely varying performance results, such as voltage output and run-times.

Lithium-ion battery cells operate at higher voltages than other rechargeable, typically about 3.6 V, versus the 1.2 V for Ni-cad, or nickel metal hydride, and 2 V for SLA. The voltage output of the battery pack is increased by adding cells that are wired in series.

The capacity is increased by adding cells in parallel. The higher voltage of lithium ion means a single cell can often be used rather than the multiple cell that are required when using older battery technology. Although, nickel metal hydride systems can be configured with up to 10 cells in a series to increase voltage, resulting in a maximum aggregate voltage of 12.5 V, lithium-ion battery systems can be configured up to 7 cells in series to increase voltage, resulting in a maximum aggregate voltage of 25 V.

Nickel metal hydride cells have 500 duty cycles per lifetime, less than 0.5C optimal load current, and an average energy density of 100 Wh/kg. They feature a less than 4-h charge time, and a typical discharge rate of approximately 30% per month while in storage.

Lithium-ion battery characteristics include a nominal voltage of 3.6 V, 500 duty cycles per lifetime, and a less than 1C optimal load current. In addition, they have an energy density of 175 Wh/kg, a less than 4-h charge time, and a typical discharge rate of approximately 10% per month while in storage.

SLA and Ni-cad batteries used to be the only options for designers of portable equipment. While these chemistries still have the advantage of lower cost and large operating temperature ranges, they have the draw back of low energy density, so products are big and heavy.

Three cell chemistries, nickel metal hydride, lithium ion, and lithium polymer, which is really a subset of lithium ion, currently dominate the growing portable markets. While all of these chemistries can address the high-power demands of portable applications, each has unique characteristics that make it appropriate, or inappropriate, for a particular portable device.

The specific characteristics of each cell chemistry, in terms of voltage, cycles, load current, energy density, charge time, and discharge rates, much be understood in order to specify a cell that is appropriate for a given application.

Lithium-ion battery systems are a good option when requirements specify lower weight, higher energy density or aggregate voltage, or a greater number of duty cycles. It offers the highest voltage per cell, a flat discharge curve, and advanced/accurate fuel gauging.

Nickel metal hydride batteries excel when lower voltage requirements, or price sensitivity, are primary considerations in cell selection. The downside is its potential for thermal runaway, temperature performance, lower energy density that Li-ion and high self-discharge rates.

Still, in some extremely cost-sensitive applications where size is not an issue and float charging is necessary, SLA is still the best choice.

Li-polymer has a lower energy density than Li-ion. However, it is safer and thinner than Li-ion. Figure 5.5[1] illustrates that lithium-ion technology offers a pronounced energy density increase. For their size and weight Li-ion cells store and deliver more energy than other rechargeable batteries. Energy density is measured both volumetrically and gravimetrically. Li-ion technology is now up to almost 500 Wh/l and 200 Wh/g. Lithium ion is able to deliver more power with a smaller footprint and less weight. The higher the energy density of the batteries used in designing a portable device, the more size is decreased and convenience is increased for portable product. Shrinking the battery pack compartment will allow the design of additional device capabilities in the same product.

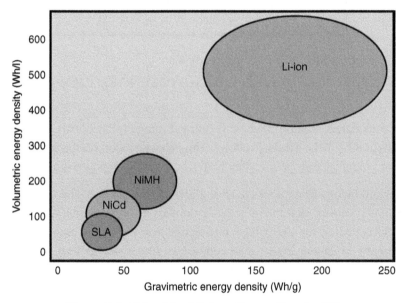

Figure 5.5: Li-ion Has Highest Energy Density [1]

Lithium-ion batteries offer many attractive advantages for portable equipment over the older rechargeable technologies. Li-ion advantages include a much higher energy density, lighter weight, longer cycle-life, superior capacity retention, and broader ambient-temperature endurance.

From the system designer's point of view, the physical properties of interest in a battery are output voltage and battery capacity. In an ideal battery, the voltage is constant over a complete discharge cycle, and it drops to zero when the battery is fully discharged. In practice, however, voltage decreases as the time of discharge increases.

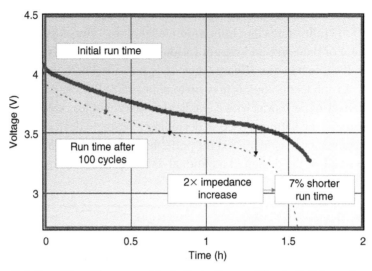

Figure 5.6: Run-Time Decreases Each Cycle Due to Rise in Cell Impedance [1]

A battery is considered exhausted when its output voltage falls below a given voltage threshold (such as 80% of the nominal voltage). This behavior motivates the adoption of DC–DC converters for voltage stabilization when batteries are used to power-up digital systems.

There are several factors that affect the Li-ion cell's performance. The first is its age. In Figure 5.6, you can see that the impedance doubles during the first 100 charge and discharge cycles, decreasing the voltage during discharge. The 2× higher impedance can affect run-time by 7% under higher load conditions. Impedance does affect the ability to get energy out of the battery systems, so design must allow for sufficient run-time even after many cycles. The battery industry defines the cell as "dead" when it degrades to 80% of its rated capacity. While the cell will still function, performance begins to degrade more rapidly. This 80% mark is often reached after about 500 cycles.

Temperature and current can also affect a battery's performance. As shown in Figure 5.7, the capacity, and therefore, the run-time of the battery can vary by 65% when the temperature is varied from 5°C to 45°C, and currents of up to 0.8 Amps. This effect can be greater or smaller depending only on slight variations in cell chemistry so different vendors' cells may perform differently in specific applications.

Figure 5.8 is an image of a battery pack with the plastic enclosure opened to see the cells and electronics. The main components of a battery pack include, the cells, the printed circuit board which provides the intelligence of the systems with features such as the

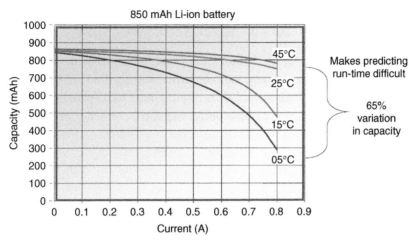

Figure 5.7: Temperature and Current Levels Can Cause Variations in Capacity [1]

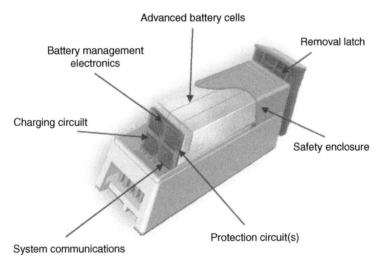

Figure 5.8: Advanced Battery Pack Features [1]

fuel gauge, the protection circuitry, the thermal sensors, any pack status indicators, and a serial data communications bus. Other components of a battery pack include the plastic enclosure, the removal latch, and any external contacts.

5.5 Portable Device Display Technologies

Display technologies are relatively new. The cathode ray tube was developed less than 100 years ago. For the last 10 years, scientists and engineers have been working closely to create a display technology capable of providing a paper and ink like reading experience, with superior viewability, but also with respect to cost, power, and ease of manufacture.

An LCD display system is composed of an LCD panel, a frame buffer memory, an LCD controller, and a backlight inverter and lamp or light-emitting diode (LED), as shown in Figure 5.9. High-resolution, high-color LCDs require large LCD panels, high-wattage backlight lamps, and large-capacity frame buffer memories, which together lead to high-power consumption.

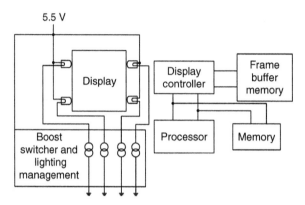

Figure 5.9: Block Diagram of a Display System

The processor and the memory are in power-down mode during the slack time, but the display components are always active mode, for as long as the display is turned on. This makes the LCD backlight the dominant power consumer, with the LCD panel and the frame buffer coming a second and third in power consumption. A modern mobile device requires a lot of computing power. With interactive applications, such as a video telephony or an assisted GPS, an even higher portion of the energy will be consumed by the display system.

5.5.1 Mobile Device Power Distribution

Figures 5.10–5.12[2] indicate pie charts that illustrate the mobile device power distribution ranging from a legacy mobile device (voice only) to a smartphone/multimedia mobile device to a gaming targeted mobile device. Convergence of features is driving new application processing and visual display requirements.

5.5.2 Backlights

Backlight mechanical design has become very sophisticated, allowing very few LEDs to be used with highly complex optical light pipes/light spreaders which create uniform

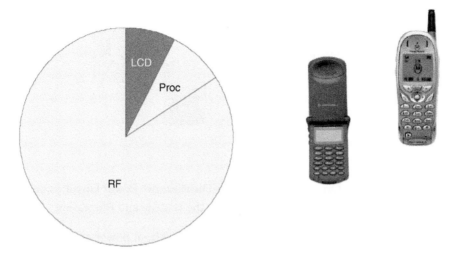

Figure 5.10: Legacy Handset Power Distribution Radio Frequency (RF) Dominated Power Consumption in Legacy (Voice Only) Handsets

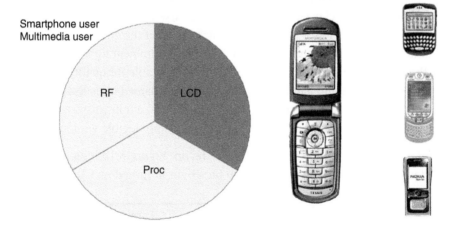

Figure 5.11: Feature Rich Handsets Power Distribution. More Equitable Power Consumption Distribution in Smartphones/Multimedia Mobile Devices

illumination of the LCD. Power consumption of the backlight is a critical issue. Now that white LEDs have replaced cold-cathode fluorescent lamps (CCFLs) for backlighting mobile device displays (and LEDs will soon replace CCFLs in laptops also), the next major innovation is the RGB LEDs, which will significantly enlarge the color gamut of the backlight, and therefore the display itself. It is claimed that an image that has higher color saturation (larger color gamut) looks brighter than a lower color saturation image.

Single-player gamer

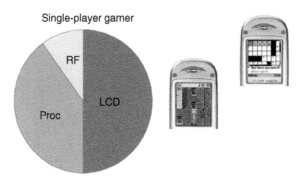

Figure 5.12: Game Oriented Phone Power Distribution. Power Distribution for Single Game Players, Dominated by the Display and Processing

Engineers will take advantage of this effect to lower backlight power consumption. The RGB LED backlights will be a great benefit to mobile device displays.

Many mobile devices are equipped with color thin-film transistor (TFT) liquid–crystal displays (LCDs). A quality LCD system is now the default configuration for handheld embedded systems. An LCD panel does not illuminate itself and thus requires a light source. A transmissive LCD uses a backlight, which is one of the greediest consumers of power of all system components. A reflective LCD uses ambient light and a reflector instead of the backlight. However, reflective LCD is not suitable for quality displays, and complementary use of the ambient light and the backlight, named transflective LCD, is used for small handheld devices. When the backlight is turned off, a transmissive LCD displays nothing but black screen; even transflective screens are barely legible without the backlight.

Most useful applications require the backlight to be on. Figure 5.13 indicates the display power modes and the current drawn when the backlight in turned on [2].

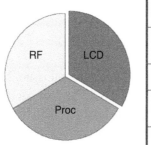

Display power modes	LCD panel	Backlight
Active transmissive	ON (5 mA)	ON (60–120 mA)
Active reflective	ON	OFF
Partially active reflective	Partially ON (<1 mA)	OFF
Inactive	OFF	OFF

Figure 5.13: Most Power Is Consumed When the Backlight Is On

Figure 5.14 indicates the system level components contribution to the power consumption. Note the power consumed by the backlight.

There are many techniques employed to conserve energy consumed by a display system. Ambient luminance affects the visibility of LCD TFT panels. However, by taking account of this, backlight autoregulation [4] can also reduce the average energy

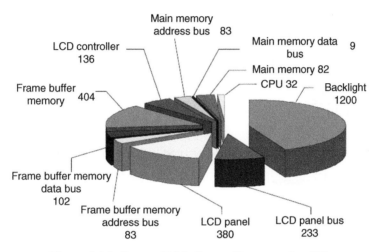

Figure 5.14: System Wide Power Consumption [3]

requirement of the backlight. Simultaneous brightness and contrast scaling [5] further enhances image fidelity with a dim backlight, and thus, permits an additional reduction in backlight power. Even more aggressive management of the backlight can be achieved by modification of the LCD panel to permit zoned backlighting [6]. Additional energy conserving techniques include dynamic luminance scaling (DLS), dynamic contrast enhancement (DCE), and backlight autoregulation. These techniques will be discussed later in this section.

5.5.3 Display Technologies

Display technologies such as backlight LCDs, reflective LCDs, electroluminescent (EL) displays, organic LEDs (OLEDs), and electrophoretic displays (EPD) objective is to achieve paper-like viewing displays.

There are four primary approaches to flat-panel displays. Three approaches are illustrated in Figure 5.15:

1. *Transmissive* displays work by modulating a source of light, such as a backlight, using an optically active material such as a liquid–crystal mixture.

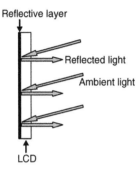

Reflective LCD: external light reflected and modulated by an optically active material

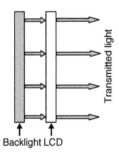

Transmissive LCD: light source modulated by an optically active material

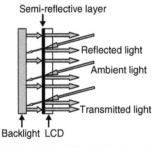

Transflective LCD: combination of transmissive and reflective diplays

Figure 5.15: Approaches to Displays

Source: www.hitachi-dispalys.eu.com

2. *Emissive* displays such as OLEDs make use of organic materials to generate light when exposed to a current source.

3. *Reflective* displays work by modulating ambient light entering the display and reflecting it off of a mirror-like surface. Until recently, this modulation has typically been accomplished using liquid–crystal mixtures or electrophoretic mixtures.

4. *Transflective* displays are a hybrid combination of a transmissive and reflective display. This technology was developed to provide sunlight viewability for transmissive displays. Being a compromise however, this type of display technology offers a compromised viewing experience.

Reflective displays were invented primarily to address the shortcomings of transmissive and emissive displays, namely power consumption and poor readability in bright environments.

Since transmissive LCDs require a power-hungry backlight and emissive OLEDs require a constant power source to generate light, it makes it difficult for designers of these technologies to reduce power consumption. This is especially important for battery-powered portable devices such as mobile phones, PDAs, digital music players, digital cameras, GPS units, and mobile gaming devices. With efficient use of ambient light, reflective displays eliminate the backlight unit and offer both significant power savings and a thinner display module (Table 5.2).

Mobile device display technologies can be separated into emissive and non emissive classes. This classification is expanded and illustrated in Figure 5.16.

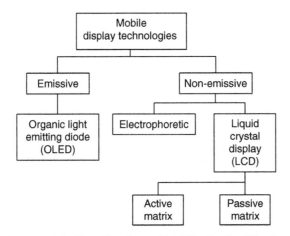

Figure 5.16: Classification of Mobile Device Displays

5.5.3.1 TFT LCD

There are three types of TFT LCD panels. In *transmissive* LCDs, a backlight illuminates the pixels from behind. Transmissive LCDs offer a wide color range and high contrast, and are typically used in laptops. They perform best under lighting conditions ranging from complete darkness to an office environment.

Reflective LCDs are illuminated from the front. Reflective LCD pixels reflect incident light originating from the ambient environment or a frontlight. Reflective LCDs can offer very low-power consumption (especially without a frontlight) and are often used in small portable devices such as handheld games, PDAs, or instrumentation. They perform best in a typical office environment or in brighter lighting. Under dim lighting, reflective LCDs require a frontlight.

Transflective LCDs are partially transmissive and partially reflective, so they can make use of environmental light or backlight. Transflective LCDs are common in devices used under a wide variety of lighting conditions, from complete darkness to sunlight.

Transmissive and transflective LCD panels use very bright backlight sources that emit more than 1,000 cd/m^2. However, the transmittance of the LCD, ρT, is relatively low, and thus the resultant maximum luminance of the panel is usually less than 10% of the backlight luminance.

Theoretically, the backlight and the ambient light are additive. However, once the backlight is turned on, a transflective LCD panel effectively operates in the transmissive mode because the backlight source is generally much brighter than the ambient light.

As stated earlier, backlighting for LCDs is the single biggest power draw in portable displays. This is especially true in bright environments where the backlight has to be switched to the brightest mode. Given how difficult it is to view a typical *transmissive* LCD in a sunlit environment, LCD developers have been working diligently on *reflective* LCDs.

Currently there are a number of portable devices using *transflective* LCDs. The transflective display was invented to improve the performance of the transmissive LCD outdoors, where bright ambient light quickly overpowered the LCD backlight, making the display hard to read. It was also configured to address the shortcomings of a purely reflective LCD in a dark environment. The transflective display employs a reflector that lets some light through from a backlight. Using such an element, the display can be used in the dark where the backlight provides illumination through the partly transmissive reflecting element. In the bright outdoors, the backlight can be switched off to conserve power and the mirrored portion of the reflector allows the LCD to be viewed by making use of the ambient light. Theoretically, the transflective display appears to fix the shortcomings of the purely reflective and transmissive displays. But in reality, this approach is a compromise and offers a rather poor viewing experience.

Figure 5.17 shows the complexity of an LCD. The extensive use of optical films such as polarizers and color filters, as well as the TFT element which itself requires several process steps to fabricate. Since LCDs work with polarized light, the necessity of using a polarizer limits the amount of light that is reflected or transmitted from the display. The additional layers, such as the color filter, reduce light even further. Consequently, today's LCDs require brighter backlights in order to be readable, whether in total darkness or in the bright sunlight. These brighter backlights lead to greater power consumption.

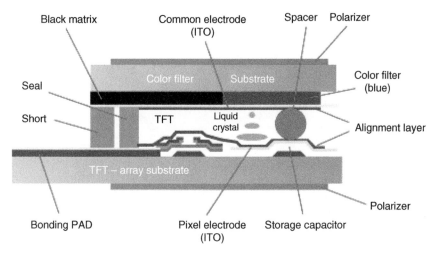

Figure 5.17: LCD Structure

Source: http://www.teac.com.au

Despite the ever increasing advances in LCD's technology, their power consumption is still one of the major limitations to mobile. There is a clear trend towards the increase of LCD size to exploit the multimedia capabilities of mobile devices that can receive and visualize high definition video and pictures. Multimedia applications running on these devices impose LCD screen sizes of 2.2–3.5 in. and more to display video sequences and pictures with the required quality.

5.5.3.2 OLED

Similar to LCDs, OLEDs can be constructed using a passive or active matrix. The basic OLED cell structure is comprised of a stack of thin organic layers that are sandwiched between a transparent anode and a metallic cathode.

When a current passes between the cathode and anode, the organic compounds emit light (see Figure 5.18). Unlike LCDs, passive matrix OLEDs does not suffer from lower contrast or slower response time. However, OLEDs offer several advantages over LCDs.

The obvious advantage is that OLEDs are like tiny light bulbs, so they do not need a backlight or any other external light source. They are less than one-third of the bulk of a typical color LCD and about half the thickness of most black-and-white LCDs. The viewing angle is also wider, about 160°. OLEDs also switch faster than LCD elements, producing a smoother animation. Once initial investments in new facilities are recouped, OLEDs can potentially compete at an equal or lower cost than incumbent LCDs.

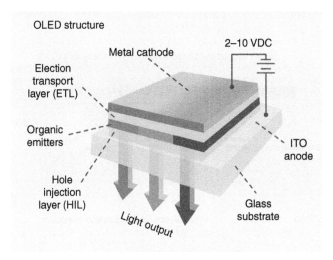

Figure 5.18: OLED Structure

Source: http://www.usami.princeton.edu

Despite these advantages, OLEDs have a relatively short lifespan and as power/brightness is increased the life is reduced dramatically. This is especially true for the blues, which lose their color balance over time. In addition, only low-resolution OLED displays can use passive matrix backplanes and higher resolutions require active matrices, which need to be highly conductive since OLEDs are current driven. Typically, low temperature poly silicon (LTPS) backplanes are used which adds cost and complexity. These conductors are also highly reflective requiring the OLED designers to add a circular polarizer on the front of the display reducing the efficiency of the display and increasing the cost. Finally, as is the case with all emissive displays, OLED displays have poor readability in environments such as the bright outdoors.

Table 5.2: Comparison of Display Technologies

		Existing Technology				
Parameter	**Units**	**Reflective STN**	**Reflective TFT**	**Backlit TFT**	**EPD**	**OLED**
Color		B&W	Color	Color	Limited color	Color
Resolution		Low	Good	Good	Good	Good
Sunlight readable	80% contrast 3000FC	Yes, good	Yes, good	No	Yes, excellent	Yes, poor
Video response speed	ms	100	20	20	500	1/1,000
Static power including drivers	mW	15	20	200	0	200
Thickness	mm	5	6	9	1.25	3
Environmental	Temperature range (°C)	10–30	10–30	10–30	0–50	10–50

5.6 Low Power LCD Techniques

A number of low-power LCD techniques have been investigated. These include DLS, extended DLS (EDLS), frame buffer compression, dynamic color depth control, variable duty ratio refresh, backlight autoregulation, and dark window optimization. Each technique saves the power consumption of the display system by reducing the activity of the corresponding components such as the backlight luminance, the color depth, the refresh duty ratio, and the pixel brightness.

5.6.1 Dynamic Luminance Scaling

DLS keeps the perceived intensity or contrast of the image as close as possible to the original while achieving significant power reduction. DLS compromises quality of image between power consumption, which fulfills a large variety of user preferences in power-aware multimedia applications. DLS saves 20–80% of power consumption of the backlight systems while keeping a reasonable amount of image quality degradation.

DLS adaptively dims the backlight with appropriate image compensation so that the user perceives similar levels of brightness and contrast with minor image distortion.

The luminance of the backlight is proportional to its power consumption. As we dim the backlight, the brightness of the image on the LCD panel is reduced, but we save power. The principle of DLS is to save power by backlight dimming while restoring the brightness of the image by appropriate image compensation [8,9].

Brute-force backlight dimming is a traditional technique to save power consumption, but it reduces brightness, and thus, the display quality is degraded. By contrast, DLS does not sacrifice the overall brightness of the image but accommodates minor color distortions. To achieve the maximum power saving for a given color distortion limit, DLS dynamically scales the luminance of the backlight as the image on the LCD panel changes.

There are a number of image compensation algorithms described in the following section.

5.6.1.1 Brightness Compensation

Brightness is the intensity of an image perceived by human eyes. As an approximation brightness is considered linearly proportional to the luminance of the LCD panel. The major computational overhead of brightness compensation is the construction of the transformation function and the transformation of each pixel color. Building the transformation function includes the construction of the histogram and determining a value for the threshold. Transformations are typically performed either by multiplication and division operations.

Brightness compensation allows a significant degree of backlight dimming while keeping the distortion ratio reasonable, as long as the image has a continuous histogram which is not severely skewed to bright areas. Although all the histograms are discrete by definition, we express that a histogram is continuous if adjacent values are similar with each other, considering the original image before digitization of the color values. Discrete histograms generally make it difficult to determine a proper threshold value and most graphical user interface (GUI) components have discrete histograms.

5.6.1.2 Image Enhancement

Image enhancement allows one to apply DLS for the images with discrete histograms where the brightest area dominates the image. Techniques of histogram stretching and histogram equalization are employed. Histogram stretching is an extension of brightness compensation, in that the histogram is stretched with respect to the low threshold as well as the high threshold. Histogram stretching truncates data not only in the brightest areas but also in the darkest areas. It generally doubles the amount of backlight dimming that can be achieved, in comparison with brightness compensation. Histogram stretching implies contrast enhancement rather than recovery of brightness. The contrast enhancement is more desirable for GUI applications, where readability is the primary objective.

Histogram stretching outperforms brightness compensation. However, building the transformation function has twice the computational complexity. Sometimes, the colors of objects are not important, but we need maximum readability. Text-based screens often fall in this category. In such cases, we need to achieve more contrast to allow more backlight dimming. Histogram equalization is a useful technique for this purpose.

Image enhancement is also applicable to images with continuous histograms. Since histogram stretching is an extension of brightness compensation, there is similar distortion to the image. The majority of pixels preserve their original colors, and thus we can apply brightness compensation and histogram stretching for streaming images where inter-frame consistency must be considered. On the other hand, histogram equalization is not applicable to streaming images because in this case most pixels change their colors. Histogram equalization has a tendency to spread the histogram of the original image so that the levels of the histogram-equalized image span a wider range [10].

Histogram equalization generally offers better readability than histogram stretching when the image has a discrete spectrum. The computational complexity for building the transformation function of histogram equalization is the same as that for brightness compensation. Since the transformation function is not a polynomial implementation, table lookup is desirable.

5.6.1.3 Context Processing

Histogram equalization generally outperforms histogram stretching in terms of readability for GUI applications, if the histogram is discrete. However, some minor colors may be merged into each other and are thus no longer distinguishable after histogram equalization. In the case of photographs, some minor colors may be merged into others or become more similar to each other. But, in the case of text, the number of pixels does not correlate with importance. So we never allow text to be merged into its background after histogram equalization.

Context processing is a useful technique to prevent small foreground objects from having similar colors to their background after histogram equalization. If a foreground color and a background color become equal or similar after histogram equalization, context processing re-stretches their colors so that the distance between them in color space is a maximum.

Context processing is a post-processing step that can be applied after brightness compensation, histogram stretching, or histogram equalization. Distortion ratio no longer has meaning if context processing is used. Context processing does not require the overhead of building a transformation function since it is not based on the histogram.

However, transformation does require context information for the application. Context processing can be implemented by addition operations only.

DLS consists of two major operations: image compensation and backlight control. Both can be implemented by modifying the application program and the operating system or the frame buffer device driver.

5.6.2 Extended DLS

DLS is extended to cope with transflective LCD panels, which operate both with and without a backlight, depending on the remaining battery energy and ambient luminance. These popular transflective LCD panels are the dominant choice for battery operated electronic systems because they allow an image to remain visible without a backlight, even though the quality can be poor.

Remember the principle of DLS is to reduce the light source's luminance but compensate for the loss in brightness by allowing more light to pass through the screen, enhancing the image luminance. The viewer should perceive little change. DCE also enhances image quality under a dimmed backlight, but does so by increasing the image's contrast. DCE requires similar image processing to DLS, and thus we have the same degree of freedom in adaptation. Although DLS preserves the original colors, DCE results in a noticeable change to the original colors in pursuit of higher contrast and improved legibility. DCE is a very aggressive power management scheme for transmissive LCD panels, which differentiates it from DLS. The EDLS framework, as illustrated in Figure 5.19, achieves a congruent combination of DLS and DCE.

Panel mode	Transmissive mode		Reflective mode
Backlight	Full	Dimmed	No backlight
Image	Original image	Brightness enhancement	Contrast enhancement
Power source	External power supply	Moderate battery power	Poor battery power

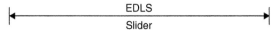

Figure 5.19: EDLS Framework [3]

Fundamentally the EDLS interface is a simple slider knob. The EDLS knob controls the trade-off between energy consumption and image quality. In addition it provides users with a power management scheme that can extend battery life at the cost of whatever Quality-of-Service (QoS) degradation the user will accept. There is also an automatic mode that changes the power management setting, depending on the remaining battery energy.

When connected to an external power source, the backlight is fully on and exhibits its maximum luminance. There should be no backlight power management so that users can enjoy the best image quality. When the system is battery powered, however, users might want to extend the battery life for future use, even if the battery is already fully charged. But users generally are not ready to sacrifice appreciable picture quality at that stage. As the remaining battery energy decreases, users might become increasingly willing to compromise image quality to extend battery life. This is the point at which EDLS applies DLS.

With a poor power budget, the user's prime concern might well be to complete the current task within the remaining battery energy budget, even if the image quality decreases. This is the optimum time for EDLS to change from DLS to DCE mode. Although DCE might alter the original colors, a moderate degree of DCE does at least maintain a fixed distortion ratio. However, if the battery energy is nearly exhausted, the only remaining option is to turn off the backlight. Without the backlight, EDLS applies DCE to achieve the maximum possible contrast. In this case, EDLS cannot guarantee a fixed amount of image distortion, but the user should still be able to read the display and finish the task.

The EDLS process starts by building a red–green–blue histogram of the image for display. The EDLS slider determines the panel mode (transmissive or reflective), the image processing algorithm (DLS or DCE), and the maximum allowed percentage of saturated pixels, SR, after image processing. EDLS process derives upper and lower thresholds TH and TL from SR and the histogram, and calculates a scaling factor that controls the amount of backlight dimming,

EDLS significantly reduces backlight power consumption. However, it results in power, delay, and area overheads that take place in other components. These overheads are primarily determined by the screen resolution, refresh rate, and color depth.

5.6.3 Backlight Autoregulation

A mobile device operating in an environment with low ambient luminance, and this luminance can be detected by a photo sensor, the backlight can be dimmed without effecting the user. Backlight autoregulation adaptively dims the backlight in response to changes in the ambient luminance [4]. Backlight autoregulation is applicable while maintaining QoS only when reduced contrast by backlight dimming does not compromise the visibility. The contrast between the LCD panel with normal backlight and the dark environment with low ambient luminance is high enough so that we can safely reduce the contrast by dimming the backlight without compromising the visibility. To take advantage of backlight autoregulation, a mobile device must be equipped with a photo sensor to detect the ambient luminance. Typically the photo sensor can be the on board camera.

5.6.4 Frame Buffer Compression

Frame buffer compression [11] is used to reduce the power consumption of a frame buffer memory and its associated buses. LCD controllers periodically refresh their display at 60 Hz activating the frame buffer.

Frame buffer compression reduces the activity of the frame buffer and thus its power consumption. The compression algorithm employed is based on run-length encoding for on-the fly lossless compression.

An adaptive and incremental re-compression scheme, to accommodate frequent partial frame buffer updates efficiently, has been developed. The result is a savings from 30% to 90% frame buffer activity on average for various mobile applications. The implementation of compression scheme consumes 30 mW more power and 10% more silicon space than a conventional LCD controller without frame buffer compression. However, the power saved in the frame buffer memory is up to 400 mW.

5.6.5 Dynamic Color Depth

Dynamic color depth control [8] modifies the pixel organization in the frame buffer, which enables half of the frame buffer memory devices to go into power-down mode at the cost of a decreased color depth.

Dynamic color depth control achieves an energy saving from the frame buffer. Variable duty ratio refresh [8] reduces the duty ratio of refresh cycles as far as possible. This occurs only if the time constant of the storage capacitor of a sub-pixel on the TFT LCD panel is higher than the refresh period, saving power in the frame buffer and the LCD panel interface bus.

Engineers have been working on backlight autoregulation [4], which adaptively dims the backlight in response to changes in the ambient luminance, and a dark window optimization [12] which modifies the windowing environment to allow changes to the brightness and color of areas of the screen that are not of current interest to the user. This saves power in OLED display panels.

There are many techniques available for low-power display systems. Choosing the proper techniques are very important. Frame buffer compression and dynamic color depth control have the same goal of reducing the power consumption from the frame buffer. However, they cannot both be applied at the same time, and so we have to select one. The user may be willing to allow some decrease in color depth in exchange for higher contrast in a document viewer, where image legibility is the most important QoS requirement, and dynamic color depth control can meet user's preferences. But if a photo image viewer is running, then image fidelity should be preserved, and we should adopt frame buffer compression.

5.7 Summary

The following is a brief summary of two of the most important components that comprise mobile devices. Batteries are the energy source and displays the largest consumer of that energy.

5.7.1 Batteries

The battery section illustrated that the evolution of batteries has been steady. However, current battery technology is failing to keep up with Moore's law – no Moore's law doubling for battery technology. In addition the demand for energy from a rechargeable battery continues to grow exponentially [13] as wireless broadband, rich multimedia technologies such as video, drive the use time between charges to an unacceptable level.

Greater operating demands create new performance challenges. Some of these challenges include:

- Cell imbalance
- Temperature effects
- High pulse currents
- Charging regimes
- Gauging accuracy
- Storage effects

Work continues on finding a future successor to lithium ion and lithium-ion polymer batteries. The goal is to make the battery cheaper and increase its energy density, while

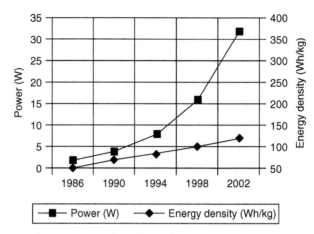

Figure 5.20: Widening of the Battery Gap [13]

increasing battery safety (Figure 5.20). The last goal is important because currently lithium batteries are required to have a protection circuit which can add $1.00–$1.25 to the cost of the battery. Eliminating this circuit has an immediate effect on the bottom-line.

Research is centering on finding a replacement for the cathode electrode in conventional lithium-ion cells. Not only is the commonly used lithium cobalt oxide very expensive, but also it accounts for much of the potential instability of a lithium-ion battery. Other cathode materials that are being explored include nickel, magnesium, and nickel silicon. Potentially these materials could lead to cells with energy densities exceeding 600 Wh/l, in addition to eliminating the need for protection circuits.

Fuel cells, discussed further in Chapter 8, are considered the future in mobile handset power sources, but concerns for their safety, high cost, operation temperature, and inconvenience remain. Fuel cells convert a fuel, usually methanol or ethanol alcohol, into energy while giving off no pollution. They are claimed to operate several times longer than a lithium-ion battery and can be re-charged with fuel in seconds.

5.7.2 Displays

The display section started with an overview of the key components that make up the display system in a mobile device. The basic block diagram of the display system comprises of controller, frame buffer memory, display, processor, and backlight components.

It was illustrated that the display system is the largest consumer of power in a mobile device, varying depending on the specific use of the mobile device from voice only to smartphones to mobile gaming devices.

In addition backlighting was identified as the component that consumes the largest quantity of power in a display system. Given this fact, a number of low-power display techniques for mobile devices have been discussed including DLS, EDLS, dynamic color depth control, backlight autoregulation, and frame buffer compression. Each of the techniques saves the power consumption of the display system by reducing the activity of the corresponding components such as the backlight luminance, the color depth, the refresh duty ratio, and the pixel brightness.

There are many techniques available for low-power display systems. However the systems developer has to choose those that are most appropriate. For example, both frame buffer compression and dynamic color depth control have the same objective of reducing the power consumption from the frame buffer. As a result they cannot both be applied at the same time, and therefore one has to be selected.

The promising goal is a framework that holistically accommodates most of the power-reduction techniques for display systems. This framework considers the dependency between several techniques, user's preferences, and the current system context. In addition it selects the most appropriate techniques to satisfy the QoS requirement as well as the power requirement.

References

[1] R. Tichey and R. Staub. Innovations in Battery and Charger Technologies for Portable Electronic Devices (Presentation), Micro Power Electronics Webinar, April 2007.

[2] T. Botzas. PenTile RGBW display technology for meeting aggressive power budgets in high resolution multimedia mobile applications. *International Wireless Industry Consortium*, Toronto, Ont., Canada, September 2005, pp. 3,5–7.

[3] I. Choi. *J2ME LBPB (Low Power Basis Profile of the Java 2 Micro Edition)*. Computer Systems Laboratory, Seoul National University, Seoul, Korea, May 2002.

[4] F. Gatti, A. Acquaviva, L. Benini, and B. Ricco. Low power control techniques for TFT LCD displays. *Proceedings of the International Conference on Compilers, Architecture, and Synthesis for Embedded Systems*, Grenoble, France, October 2002, pp. 218–224.

[5] W. C. Cheng, Y. Hou, and M. Pedram. Power minimization in a backlit TFT–LCD display by concurrent brightness and contrast scaling. *Proceedings of Design, Automation and Test in Europe*, Paris, France, February 2004, pp. 16–20.

[6] J. Flinn and M. Satyanarayanan. Energy-aware adaptation for mobile applications. *Proceedings of the Symposium on Operating Systems Principles*, Charleston, South Carolina, 1999, pp. 48–63.

[7] K. Suzuki. Past and future technologies of information displays, *Electron Devices Meeting, 2005. IEDM Technical Digest. IEEE International*, 5–7 December 2005, pp. 16–27.

[8] I. Choi, H. Shim, and N. Chang. Low-power color TFT LCD display for hand-held embedded systems. *Proceedings of the International Symposium on Low-Power Electronics and Design*, Monterey, CA, USA, August 2002, pp. 112–117.

[9] I. Choi, H. S. Kim, H. Shin, and N. Chang. LPBP: low-power basis profile of the Java 2 micro edition. *Proceedings of the International Symposium on Low-Power Electronics and Design*, Seoul, Korea, August 2003, pp. 36–39.

[10] R.C. Gonzalez and R.E. Woods. *Digital Image Processing*, 3rd ed. Reading, MA: Addison-Wesley, 1992.

[11] H. Shim, N. Chang, and M. Pedram. A compressed frame buffer to reduce display power consumption in mobile systems. *Proceedings of ACM/IEEE Asia South Pacific Design Automation Conference*, Yokohama, January 2004, pp. 819–824.

[12] S. Iyer, R. Mayo, and P. Ranganathan. Energy-adaptive display system designs for future mobile environments. *USENIX Association Proceedings of MobiSys 2003: The First International Conference on Mobile Systems, Applications and Services*, San Francisco, CA, May 2003.

[13] K. Lahiri, A. Raghunathan, S. Dey, and D. Panigrahi. Battery-driven system design: a new frontier in low power design. *VLSI Design/ASPDAC'02*, Banglore, India, January 2002.

CHAPTER 6

Power Management Integrated Circuits

6.1 Introduction

Power management needs have proliferated exponentially with the variety of mobile phones, and device features and functions. As long as the mobile phone was simply required to make and receive phone calls, it embodied one set of power management requirements. However, as mobile phone manufacturers have heaped on the features, each requirement has placed another demand on power management.

Raising the data rate for mobile phones and adding an Internet browser put extra demands on battery life. Color liquid–crystal display (LCD) screens with white LED backlighting taxes battery life, as does the complementary metal-oxide-semiconductor (CMOS) image sensors used to take pictures for camera phones. New embodiments of the camera phone compete with digital still cameras, with mega pixel sensors, auto focus motors, and high-intensity white LED flash attachments. The same portable device may include an MPEG-4 codec for video and an MP3 audio player, with built-in piezo-electric speakers and surround-sound decoders.

In addition to this, the mobile phone must also work in different areas of the world. It must include multiple radio transmitters to respond to different standards. To enable communication between PC databases and hands-free headsets, the mobile device will include Bluetooth and 802.11 Wireless LAN attachments.

Power management would be much simpler if all portable electronics required the same operating voltage. But typical mobile devices use a number of voltages. For example, one voltage is required to power the mobile device's central processor, one for the core, another for the I/O sections, and one for the backlight for the display. At full power, these regulators can output 400–600 mA of current.

The driver circuitry for speakers, headphones, displays, and radio frequency (RF) transceiver components each have different voltage and current requirements. Even where the supply voltage requirements are the same, separate voltage regulators buffer the drivers from the baseband processor and from each other. The regulator for a brightly lit display, for example, would prevent a voltage drop on its own line from affecting the performance of other integrated circuit (IC) components.

A diagram of evolving mobile phone features and functions is shown in Figure 6.1. How anyone gets more than a couple of minutes of battery life out of such a device is among the wonders of modern electronics. Some mobile phones incorporate micro-miniature hard disk drives to store audio, pictures, and video. This represents an additional drain on battery life.

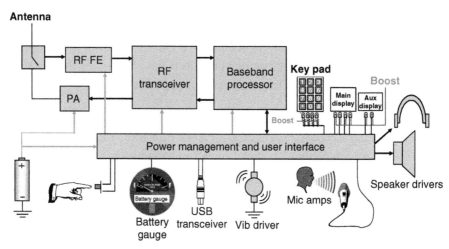

Figure 6.1: Supporting the Evolving Features of Wireless Mobile Devices

Source: www.freescale.com

The SoCs required to implement features, like small LCDs with white LED backlights, silicon imagers and camera flash attachments, high-fidelity audio codecs and headphone amplifiers, require specialized voltages from the mobile handset battery source. For instance, in Japan, where schoolchildren use high-fidelity ring tones to personalize their mobile handsets, the handsets must include power-consuming piezo-electric speakers and drivers. In all cases, a series of voltage regulators must convert the 3.6 V available from a typical Li-ion handset battery to the 3.3, 3.0, 2.8, 2.5, 1.8, or 1.5 V needed to power the circuitry.

The voltage regulators must be highly efficient to minimize any battery power consumption of their own. They also must serve a management function for the particular devices they power, putting those devices in a quiescent mode when they are not in use. Thus, a feature-laden mobile handset may embody 5–15 separate voltage regulators and power monitors, as well as diodes, transistor switches, and sequencers.

At the heart of most wireless handsets beats a power management integrated circuit (PMIC). The PMIC handles most of the power-supply requirements and other blocks including:

- Linear regulators
- Switching regulators
- Battery charger
- Battery protection unit
- Battery authentication
- Audio amplifiers
- Display drivers
- Control logic
- Reference bias generator
- Data conversion
- Connectivity
- Lighting

Leading analog semiconductor manufacturers provide PMICs [1] as full custom, semi custom, and/or standard products. Any block that is not already integrated elsewhere in the handset is a candidate for integration in the PMIC.

6.2 Voltage Regulators

Every electronic circuit is designed to be powered by some supply voltage, which is usually assumed to be constant. As the battery output voltage goes down over time a voltage regulator is necessary to maintain the required system voltage. In addition, the device must provide tight output voltage regulation. A voltage regulator provides this constant DC output voltage and contains circuitry that continuously holds the output voltage at the design value regardless of changes in load current or input voltage.

A linear regulator operates by using a voltage-controlled current source to force a fixed voltage to appear at the regulator output terminal (see Figure 6.2).

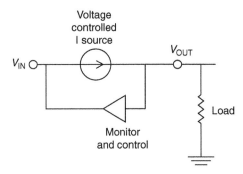

Figure 6.2: Linear Regulator

The control circuitry must sense the output voltage and adjust the current source to hold the output voltage at the desired value. The design limit of the current source defines the maximum load current the regulator can source and still maintain regulation.

The output voltage is controlled using a feedback loop, which requires some type of compensation to assure loop stability. Most linear regulators have built-in compensation, and are completely stable without external components.

In addition, a linear regulator requires a finite amount of time to adjust the output voltage after a change in load current demand. This "delay" defines the characteristic called transient response, which is a measure of how fast the regulator returns to steady-state conditions after a load change.

6.2.1 Control Loop Operation

The operation of the control loop in a typical linear regulator will be detailed using the simplified schematic diagram in Figure 6.3.

The pass device in this regulator is made up of an NPN Darlington driven by a PNP transistor. The current flowing out of the emitter of the pass transistor is controlled by Q_2 and the voltage error amplifier. The current through the R_1, R_2 resistive divider is assumed to be negligible compared to the load current.

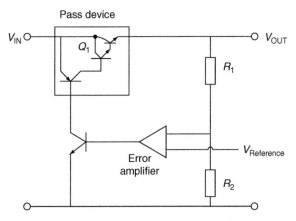

Figure 6.3: Control Loop Operation

The feedback loop which controls the output voltage is obtained by using R_1 and R_2 to "sense" the output voltage, and applying this sensed voltage to the inverting input of the voltage error amplifier. The non-inverting input is tied to a reference voltage, which means the error amplifier will constantly adjust its output voltage to force the voltages at its inputs to be equal.

The feedback loop action continuously holds the regulated output at a fixed value which is a multiple of the reference voltage (as set by R_1 and R_2), regardless of changes in load current.

It is important to note that a sudden increase or decrease in load current demand, a "step" change in load resistance, will cause the output voltage to change until the loop can correct and stabilize to the new level. This is called a transient response.

The output voltage change is sensed through R_1 and R_2 and appears as an "error signal" at the input of the error amplifier, causing it to correct the current through Q_1.

6.2.2 Linear Regulators

There are three basic types of linear regulator designs:

1. Standard (NPN Darlington) regulator

2. Low dropout or LDO regulator

3. Quasi-LDO regulator

The single most important difference between these three types is the dropout voltage, which is defined as the minimum voltage drop required across the regulator to maintain output voltage regulation. The LDO requires the least voltage across it, while the standard regulator requires the most voltage.

A critical point to be considered is that the linear regulator that operates with the smallest voltage across it dissipates the least internal power and has the highest efficiency.

6.2.2.1 The LDO Regulator

An LDO is a type of linear regulator [2, 3]. A linear regulator uses a transistor or field effect transistor (FET), operating in its linear region, to subtract excess voltage from the applied input voltage, producing a regulated output voltage. Dropout voltage is the minimum input to output voltage differential required for the regulator to sustain an output voltage within 100 mV of its nominal value.

Low dropout refers to the smallest difference between the input and output voltages that allow the LDO IC to still regulate the output voltage. That is, the LDO device regulates the output voltage until its input and output approach each other at the dropout voltage.

Ideally, the dropout voltage should be as low as possible to minimize power dissipation and maximize efficiency. And because of this LDO voltage, the LDO extends battery life by permitting the battery to be discharged all the way down to a few hundred millivolts of the desired output voltage.

The LDO's main components are a power semiconductor (pass transistor), error amplifier, and voltage reference (see Figure 6.4). One input to the error amplifier, set by resistors R_1 and R_2, monitors a percentage of the output. The other input is a stable voltage reference (V_{REF}). If the output voltage increases relative to V_{REF}, the error amplifier changes the pass-transistor's output to maintain a constant output voltage (V_{OUT}).

The LDO regulator differs from the standard regulator in that the pass device of the LDO is made up of only a single PNP transistor.

The minimum voltage drop required across the LDO regulator to maintain regulation is just the voltage across the PNP transistor:

$$V_{D(min)} = V_{CE}$$

LDO regulators, for positive-output voltages, often use a PNP for the power transistor (a pass device). This transistor is allowed to saturate, so the regulator can have a very low

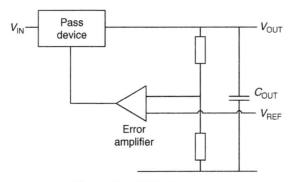

Figure 6.4: LDO Regulator

drop-out voltage, typically around 200 mV compared with around 2 V for traditional linear regulators using an NPN composite power transistor. A negative-output LDO uses an NPN for its pass device, operating in a manner similar to that of the positive-output LDO's PNP device. Developments using a CMOS power transistor can provide the lowest dropout voltage. With CMOS the only voltage drop across the regulator is the on-resistance of the power device times the load current. With light loads this can become just a few tens of millivolts.

The dropout voltage is directly related to load current, which means that at very low values of load current the dropout voltage may be as little as 50 mV. The LDO regulator has the lowest (best) dropout voltage specification of the three regulator types.

The lower drop-out voltage is the reason LDO regulators dominate battery-powered applications, since they maximize the utilization of the available input voltage and can operate with higher efficiency. The explosive growth of battery-powered consumer products in recent years has driven development in the LDO regulator product line.

The LDO regulator is best suited for battery-powered applications, because the LDO voltage translates directly into cost savings by reducing the number of battery cells required to provide a regulated output voltage. If the input–output voltage differential is low the LDO is more efficient than a standard regulator because of reduced power dissipation resulting from the load current multiplied times the input–output voltage differential.

A regulator's dropout voltage determines the lowest usable input supply voltage. Specifications may indicate a broad input-voltage range; the input voltage must be greater than the dropout voltage plus the output voltage. For a 200 mV dropout LDO, the input voltage must be above 3.5 V to produce a 3.3 V output.

With an LDO, the difference between input voltage and output voltage must be small, and the output voltage must be tightly regulated. In addition, transient response must be fast enough to handle loads that can go from zero to tens of milleamperes in nanoseconds. Also, output voltage can vary due to changes in input voltage, output load current, and temperature. Primarily, these output variations are caused by the effects of temperature on LDO voltage reference, error amplifier, and its sampling resistors R_1 and R_2.

LDOs provide either an adjustable or fixed-output voltage. Fixed-output types exhibit output voltage variation anywhere from $\pm2\%$ to $\pm6\%$ and provide 1–5 V outputs. Adjustable-output LDOs usually offer a $\pm50\%$ variation around the nominal voltage. Some LDO families give a full range of outputs in 100 or 50 mV steps, for example, 2–6 V in 100 mV steps. This wide range of output voltages is made possible by laser-trimming the ICs during their manufacture.

While the use of LDOs typically depends on the architecture of the mobile phone, the number of slots available for low-voltage regulators varies according to the architecture of the mobile phone, particularly in relation to the CPU and peripheral sets.

6.2.3 Switching Regulators

The switching regulator is increasing in popularity because it offers the advantages of higher power conversion efficiency and multiple output voltages of different polarities generated from a single input voltage.

Mobile phones utilizing a 3.6 V lithium-ion battery depend on step-down regulators to provide the 3.3 V (or lower voltages) required by system logic and baseband processors. Switching regulators here are typically architected as "buck regulators". However, there are certain devices within the mobile phone, particularly the white LEDs, which require the voltage level to be stepped up, driven by a "boost regulator." While the white LEDs are used for back lighting in mobile phones, variability in the LED manufacturing process has required voltage regulator makers to provide "constant current" outputs for their LED drivers. This ensures there are no dim spots on an LED backlight array.

6.2.3.1 Principles of Switching: Law of Inductance

If a voltage is forced across an inductor, a current will flow through that inductor and this current will vary with time. Note that the current flowing in an inductor will be time-varying even if the forcing voltage is constant. It is equally correct to say that if a time-varying current is forced to flow in an inductor, a voltage across the inductor will result.

The fundamental law that defines the relationship between the voltage and current in an inductor is given by the equation:

$$v = L \, (di/dt)$$

Two important characteristics of an inductor that follow directly from the law of inductance are:

1. A voltage across an inductor results only from a current that changes with time. A steady DC current flowing in an inductor causes practically no voltage across the inductor.

2. A current flowing in an inductor can not change value instantly, as this would require infinite voltage to force it to happen. However, the faster the current is changed in an inductor, the larger the resulting voltage.

The principles of inductance are illustrated by the information contained in Figure 6.5.

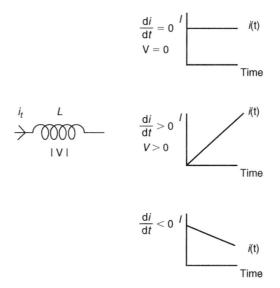

Figure 6.5: Principles of Inductance

The key parameter is the di/dt term, which is simply a measure of how the current changes with time. When the current is plotted versus time, the value of di/dt is defined as the slope of the current plot at any given point. The graph on the top shows that current, which is constant with time, has a di/dt value of zero, and results in no voltage across the inductor.

The center graph shows that a current which is increasing with time has a positive di/dt value, resulting in a positive inductor voltage. Current that decreases with time (shown in the bottom graph) gives a negative value for di/dt and inductor voltage. It is important to note that a linear current ramp in an inductor (either up or down) occurs only when it has a constant voltage across it.

Pulse-Width Modulation

All of the switching converters that will be covered in this chapter use a form of output voltage regulation known as pulse width modulation (PWM). Put simply, the feedback loop adjusts (corrects) the output voltage by changing the ON time of the switching element in the converter.

As an example of how PWM works, we examine the result of applying a series of square wave pulses to an L-C filter (see Figure 6.6).

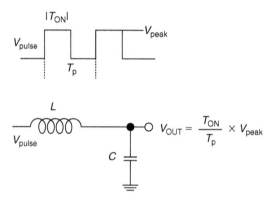

Figure 6.6: How PWM Works

The series of square wave pulses is filtered and provides a DC output voltage that is equal to the peak pulse amplitude multiplied times the duty cycle. This relationship explains how the output voltage can be directly controlled by changing the ON time of the switch.

6.2.3.2 Switching Regulator Topologies

The most commonly used DC–DC converter circuits, the Buck and Boost regulators, will now be described along with the basic principles of operation [4, 5].

6.2.3.3 Buck Regulator

IC operating voltages are decreasing relative to battery output voltages. A single Li-ion battery output can reach 4.2 V, and three new alkaline cells can produce 4.5 V. Typical required IC operating voltages are often about 3.3–1 V. A Buck converter, or step-down voltage regulator, provides switch-mode DC–DC conversion with the advantages of simplicity and low cost.

The most commonly used switching converter is the buck, which is used to down-convert a DC voltage to a lower DC voltage of the same polarity. The buck converter uses a transistor as a switch that alternately connects and disconnects the input voltage to an inductor (see Figure 6.7).

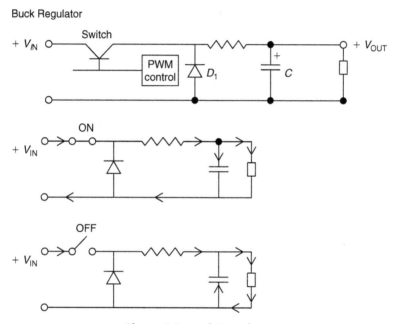

Figure 6.7: Buck Regulator

The lower diagrams show the current flow paths when the switch is on and off. When the switch turns on, the input voltage is connected to the inductor. The difference between the input and output voltages is then forced across the inductor, causing current through the inductor to increase. During the on time, the inductor current flows into both the load and the output capacitor. The capacitor charges during this time.

When the switch is turned off, the input voltage applied to the inductor is removed. However, since the current in an inductor can not change instantly, the voltage across the inductor will adjust to hold the current constant. The input end of the inductor is forced negative in voltage by the decreasing current, eventually reaching the point where the diode is turned on. The inductor current then flows through the load and back through the diode. The capacitor discharges into the load during the off time, contributing to the total current being supplied to the load. The total load current during the switch off time is the sum of the inductor and capacitor current. The shape of the current flowing in the inductor is similar to Figure 6.8.

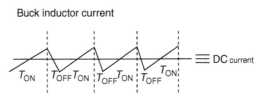

Figure 6.8: Buck Regulator Inductor Current

As explained, the current through the inductor ramps up when the switch is on, and ramps down when the switch is off. The DC load current from the regulated output is the average value of the inductor current. The peak-to-peak difference in the inductor current waveform is referred to as the inductor ripple current, and the inductor is typically selected large enough to keep this ripple current less than 20–30% of the rated DC current.

6.2.3.4 Continuous versus Discontinuous Operation

In most buck regulator applications, the inductor current never drops to zero during full-load operation (this is defined as continuous mode operation). Overall performance is usually better using continuous mode, and it allows maximum output power to be obtained from a given input voltage and switch current rating.

In applications where the maximum load current is fairly low, it can be advantageous to design for discontinuous mode operation. In these cases, operating in discontinuous mode can result in a smaller overall converter size as a smaller inductor can be used.

Discontinuous mode operation at lower load current values is generally harmless, and even converters designed for continuous mode operation at full load will become discontinuous as the load current is decreased (usually causing no problems).

6.2.3.5 Simplified Buck Regulator

Figure 6.9 shows a simplified non-isolated buck regulator that accepts a DC input and uses PWM of switching frequency to control the output of an internal power MOSFET (Q_1). An external Schottky rectifier diode, together with external inductor and output capacitors, produces the regulated DC output. The regulator IC compares a portion of the rectified DC output with a voltage reference (V_{REF}) and varies the PWM duty cycle to maintain a constant DC output voltage. If the output voltage wants to increase, the PWM lowers its duty cycle to reduce the regulated output, keeping it at its proper voltage level. Conversely, if the output voltage tends to go down, the feedback causes the PWM duty cycle to increase and maintain the proper output.

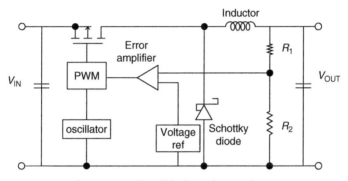

Figure 6.9: Simplified Buck Regulator

Switching frequency affects the physical size and value of external filter inductors and capacitors. The higher the switching frequency, the smaller the physical size and component value. However, there is an upper frequency limit where either magnetic losses in the inductor or switching losses in the regulator IC or power MOSFET reduce efficiency to an impractical level.

There are three ways to control the switching frequency of a regulator IC. One approach is with an external resistor. Another approach is synchronization with an external oscillator. The third approach allows frequency to be changed by connecting a pin to either ground or the VCC supply line. Not all regulator ICs provide these capabilities.

The design criteria for the component parts of a buck regulator include:

Error amplifier: For fast-response and tight regulation, the error amplifier should have a high gain-bandwidth product, preferably above 1 MHz. Some regulator ICs allow external passive components to control this, while others depend on internal parts.

Input capacitor: Select the input capacitor according to suggestions in the IC manufacturer's data sheet. Ceramic capacitors types are preferred. A low-ESR (equivalent series resistance) ceramic capacitor provides the best noise filtering of the input voltage spikes caused by rapidly changing input current. Place the input capacitor as close as possible to the IC's V_{IN} pin. In some cases, a larger value could improve filtering.

Output capacitors: Output filter capacitors smooth current flow from the inductor to the load, which helps to maintain a steady output voltage during transient load changes.

The output capacitor also aids in reducing output ripple. To perform these functions the capacitors chosen must have sufficient value and low-ESR.

Output inductor: Choose an inductor that does not saturate at the rated output current. Saturation current ratings are usually specified at 25°C, so ask the manufacturer for ratings at the maximum application temperature. Also, make sure the inductor current ripple is low enough to achieve the required output voltage ripple.

Typical buck regulator IC includes:

- Fixed- or adjustable-output voltage; some ICs have both capabilities.

- Single-ended or synchronous rectifier outputs.

- Soft-start that causes the output to come up gradually and limit inrush current.

- Power-Good output indicates if the output voltage is in regulation.

- Under-voltage lockout (UVLO) that prevents operation if the input voltage is too low.

- Thermal shutdown that cuts off regulator operation if the IC exceeds a specific temperature threshold.

- Over-current protection that deactivates the regulator if the load current exceeds a specific threshold. (This feature requires the regulator to sense its output current.)

- Over-voltage protection that prevents regulator operation if the output voltage exceeds a specific threshold.

6.2.3.6 Application

The typical Global System for Mobile Communication (GSM) handset's transmit power amplifier (PA) is powered directly from the battery for less than optimum efficiency.

In 3G handsets, the transmission of high-speed data will require increased bandwidth and increased power at the antenna; therefore, a more efficient solution is required to maintain long battery life. One architecture now gaining widespread favor among cell-phone manufacturers uses a highly specialized step-down DC–DC switching regulator to power the PA.

The principle behind the use of a switching regulator is that the PA's supply voltage headroom can be dynamically adjusted to barely accommodate the RF signal amplitude in the PA. By efficiently doing this with a switching regulator, the battery power savings are greatest when operating at anything less than peak transmit power.

Since peak power is only needed when the handset is very far away from the base station and/or transmitting data, the overall power savings are tremendous. If the PA supply voltage can be efficiently varied over a wide enough dynamic range, then a fixed-gain linear PA may be utilized. This negates the need for a separate bias control signal as presently used in 2G phones.

A bias control signal may still be utilized for an added degree of control, and some cell-phone manufacturers are actively pursuing this topology. Another major consideration to system performance is the specialized characteristics required of the step-down switching regulator. To understand the requirement, the PA must first be studied in terms of its load profile.

6.2.3.7 Boost Regulator

The boost regulator takes a DC input voltage and produces a DC output voltage that is higher in value than the input (but of the same polarity). The boost regulator is shown in Figure 6.10, along with details showing the path of current flow during the switch on and off time.

Whenever the switch is on, the input voltage is forced across the inductor which causes the current through it to increase (ramp up). When the switch is off, the decreasing inductor current forces the "switch" end of the inductor to swing positive. This forward biases the diode, allowing the capacitor to charge up to a voltage that is higher than the input voltage. During steady-state operation, the inductor current flows into both the output capacitor and the load during the switch off time. When the switch is on, the load current is supplied only by the capacitor.

Output Current and Load Power

An important design consideration in the boost regulator is that the output load current and the switch current are not equal, and the maximum available load current is always less than the current rating of the switch transistor.

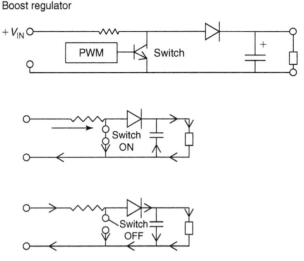

Figure 6.10: Boost Regulator

It should be noted that the maximum total power available for conversion in any regulator is equal to the input voltage multiplied times the maximum average input current (which is less than the current rating of the switch transistor).

Since the output voltage of the boost is higher than the input voltage, it follows that the output current must be lower than the input current.

6.2.4 Linear versus Switched

After looking at the specifics of linear and switched regulators, it is worthwhile making a comparison between the linear and switching types. In some cases, switching regulators can provide major benefits in a portable design. For example, if a high-performance switch-mode converter exhibits 90% efficiency, in transforming battery power to system power. However, a linear design is unlikely to extend the battery life, unless the voltage differential between the battery and linear regulator is small.

Furthermore, the linear regulator can only step a voltage down to a lower level. If a system requires voltages not available from the battery, such as high voltage for a display or a negative voltage for analog circuitry, then the system usually requires switching regulators. Table 6.1 outlines the basic differences between linear and switch-mode regulators.

There is generally a big advantage using linear regulators instead of switchers when it comes to simplicity and cost, but not efficiency. The switch-mode regulator can provide efficiencies of 90% or greater. In contrast, the linear LDO regulator usually exhibits only

Table 6.1: Differences Between Linear and Switching Regulators

Function	Linear	Switching
	Only steps down; input voltage must be greater that output	Steps up, steps down, or inverts
Efficiency	Low to medium, but actual battery life depends on load current and battery voltage over time; high if V_{IN}–V_{OUT} difference is small	High, except at very low currents (uA), where switch mode quiescent current is usually higher. Should be used for higher power load
Waste heat	High thermal dissipation if average load and/or input/output voltage difference are high	Low thermal dissipation as components usually run cool for power levels below 10 W
Complexity	Low, which usually requires only the regulator and low-value bypass capacitors	Medium to high, which usually requires inductor, diode, and filter caps in addition to the IC; for high-power circuits, external FETs are needed. Boost, buck, and combo technologies possible
Size	Small to medium in portable designs, but may be larger if heat sinking is needed	Larger than linear at low power, but smaller at power levels for which linear requires a heat sink
Total cost	Low	Medium to high, largely due to external components
Ripple/noise	Low; no ripple, low noise, better noise rejection. Not a spectral offender in radio systems	Medium to high, due to ripple at switching rate. Need to be careful with SoC and device board layout and external components

50–60% efficiency (unless V_{out} is approximately equal to V_{in}). Higher efficiency means longer battery run time.

However, the actual effect of measured efficiency on battery life can be deceptive. For many of the battery configurations, linear-regulator efficiency is quite adequate when considered over the battery's full discharge cycle.

For very-low-power designs, even a large penalty in efficiency can be acceptable. In a handheld terminal, for example, a switching supply can be worth the cost if it increases battery life from 10 to 15 days. For a small organizer, however, a similar expense just to increase battery life from 4 to 6 months may not be justified.

The major advantage of an LDO IC is its relatively "quiet" operation because it does not involve switching. In contrast, a switch-mode regulator, because they are built around high-frequency pulse generators that operate between 50 kHz and 1 MHz, can produce EMI that affects analog or RF circuits. Linear regulators generate much less ripple noise. Consequently, LDOs have been successfully deployed in mobile phones (and other low-voltage applications).

In addition, LDOs with an internal power MOSFET or bipolar transistor can provide outputs in the 50–500 mA range. The LDO's voltage and low quiescent current makes it a good fit for portable and wireless applications.

6.2.4.1 Combining Linear and Switch-Mode regulators

Over the past decade, voltage regulator makers have argued about the merits of a "buck-boost" topology for mobile phones. The buck-boost regulator would track the discharge curve of the mobile phone battery as it drifted in use from 3.6 to 2.8 V or lower. A 3.3 V buck topology would step down the battery voltage when it rose above 3.3 V. The same converter would turn into a boost regulator, elevating the supply voltage back up to 3.3 V, when the mobile phone's battery drifted below that level.

While some manufacturers were developing dual buck-boost regulator types, others argued that there was very little useful energy left in the battery when it dropped below 3.3 V and a buck-boost would itself be too expensive and consume too much power to retrieve that energy. Today, with the mobile phone laden with power-consuming features, regulator makers are playing every card they have in an effort to preserve battery life. And the buck-boost topology is back in vogue.

Combining linear- and switch-mode regulators is a common technique for generating multiple supply voltages (Figure 6.11). A linear regulator converts battery voltage to a logic supply, and one or more switchers generate other voltages required for analog circuitry and LCD-display bias.

6.2.4.2 Will a Linear Regulator Suffice?

Linear regulators are preferred in most designs. Compared with switching regulators, they provide lower cost, fewer external components, and less circuit complexity. However, linear regulators have drawbacks: reduced battery life, higher cell count, larger dropout voltage, and heat. Though not unique to portable equipment, these problems call for solutions different from those associated with AC-powered equipment.

Cell count is often an inflexible issue in determining regulator type (or vice versa). Linear regulators, for example, require a sufficient number of series-connected cells to produce

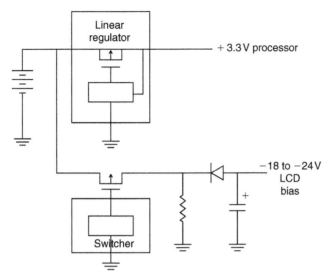

Figure 6.11: Combining Linear and Switch Mode Regulators

inputs that exceed the regulator's output voltage at all times. For a +3.3 V output, this means using three or more cells (of roughly +1 to +1.5 V each) for alkaline, NiCd, or Nimh batteries. Li+ batteries require fewer cells because Li+ cells have a higher voltage: usually between +2.5 and +4.2 V. For +5 V outputs, at least five cells may be needed to ensure a sufficient regulator input, as the cell voltages decline during discharge. For +12 V outputs, the cell count becomes so high that a switch-mode boost converter often makes more sense than a linear regulator.

Linear regulators are most appropriate when the cell count is justified from the standpoint of both voltage headroom and total energy. It is less sensible to satisfy a linear regulator's input requirement by stacking five or six cells if only two cells have enough power to support the load for an adequate time. In that case, the added cost of a switch-mode boost converter can be justified by the lower cell count, particularly if the cells are rechargeable.

If the terminal voltage of a battery falls below the desired minimum, a linear regulator cannot extract all of the energy available as the battery nears its end of discharge. A switching regulator, on the other hand, can boost the battery voltage as required. But rather than incur the expense of a switcher, the designer often selects a linear regulator with the lowest available dropout voltage.

6.2.4.3 Load Management

To reduce battery drain, many portable systems turn on their various internal circuit blocks only as needed. This switching is often implemented with logic-driven pFET switches following the regulated supply. To avoid losing regulation while delivering peak load currents, the FETs' on-resistance must be sufficiently low to ensure that the load-side voltage remains above the minimum level specified.

This switch-resistance problem is further complicated in low-voltage systems of +3.3 V and below, because the low gate drive may not sufficiently minimize the FET's on-resistance. The cost of low-gate-threshold FETs is declining. However, in many cases, the use of multiple linear regulators offers a better approach. Many new linear regulators have a logic-level shutdown capability that turns off the regulator output completely, enabling the device to serve both as a regulator and a switch.

6.2.4.4 Selecting a Regulator for Low Power

The regulator selection criteria for reduced power consumption include [6]:

- Select a regulator with LDO voltage. The lower the dropout voltage, the lower the usable input–output voltage. This reduces regulator power consumption and improves battery run time.

- Employ a regulator IC with a synchronous rectifier output stage because it's more efficient than the other possibility, an external Schottky rectifier.

- Use a switch-mode regulator that varies its switching frequency at light loads. Some regulators go into "skip" mode at light loads – that is, they minimize switching cycles at light loads. Such a regulator IC normally operates with PWM but runs at a reduced switching frequency in PFM (pulse frequency modulation) mode when experiencing light loads.

- Choose a regulator IC with a shutdown mode that disables it, cutting battery drain when the equipment is in standby. This can be done in many systems that have a normal "sleep" mode. When the IC comes out of the shutdown mode, it must do so without a transient pulse that upsets the system.

In addition, many battery-based regulator ICs have UVLO that shuts down the regulator if the input voltage drops below a specific threshold. The UVLO feature has two thresholds, one to start operation and a lower voltage to stop operation.

6.3 Battery Management: Fuel Gauges, Charging, Authentication

In addition to specialized voltage regulators for baseband processors, I/O components, LCDs, and silicon imagers, the mobile handset also requires specialized battery monitoring and charging circuits. This can be referred to as battery management [7] and is shown in Figure 6.12.

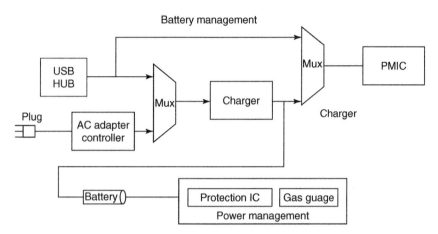

Figure 6.12: Battery Management

There are four main components to battery management:

1. *Battery charge management*: Battery charge management ICs integrate all functionality required to safely charge rechargeable batteries to maximize capacity and minimize charge time.

2. *Battery fuel gauging*: Battery fuel gauge ICs allow the system to acquire the exact state of battery charge. The information from a battery fuel gauge IC can form the basis of a complete power management routine in a portable system, enabling the system to extend its run time.

3. *Battery protection*: Li-ion battery packs need protection from overcharge or over discharge conditions. Commercial solutions offer protection for 1–4 series Li-ion battery packs. In some cases, a Li-ion protection IC can work with a battery fuel gauge IC to offer a comprehensive battery fuel gauge/protection solution.

4. *Battery authentication*: Manufacturers specify products to achieve required performance and safety goals. Authentication ensures that connected devices fulfill the requirements and are safe for the user.

The fuel gauge is one type of power management device that is finding its way into mobile devices. Fuel gauges monitor the discharge curve for rechargeable batteries.

In addition to providing an accurate indication of how much charge is left on the battery and report on the remaining power level in a useful way to the mobile handset user. Battery managers also control the charging process. The monitor is often coupled to a charging circuit to monitor battery charging levels and temperatures.

Lithium-ion batteries, which offer among the highest densities available to consumers, are potentially explosive. Lithium-ion battery chargers thus perform a safety function, preventing the battery pack from overheating during a fast-charge operation.

6.3.1 Fuel Gauges

Traditional fuel gauges monitored the voltage or the capacity, and the accuracy was quite limited. This strategy is inherently flawed for Li-ion because of the normally advantageous flat discharge profile. A new gas gauge monitors the number of Coulombs being transferred, then predicts the amount of remaining capacity, and opportunistically calibrates with the open circuit voltage of the Li-ion pack [8]. This is also a flawed strategy, because the amount of capacity changes over a battery's lifetime. Learn cycles are necessary to keep the prediction accurate. However, these features allow the end user to intelligently manage the device use and avoid unexpected failures or shutdowns.

New, accurate fuel gauges take load and environment information, including average power usage, recent power usage, and the temperature profile, and battery related information including the open circuit voltage, full charge capacity, resistance, and starting depth of discharge. And a fuel gauge performs calculations using Coulomb counting and a simplified DC-model in the case of constant current or a stepwise calculation in the case of constant power. The resulting capacity and runtime predictions are 99% accurate without the inconvenience of learn cycles.

Because the discharge curve for Li-ion batteries is non linear, specialized "fuel gauge" ICs are required. Lithium-based batteries have been known to become highly unstable in the presence of moisture. Modern sealing techniques have practically eliminated this issue for Li-ion batteries, but the charging and monitor circuits ensure that electrolytes do not overheat and damage the battery's seal.

Accurate fuel gauging, combined with smart charging algorithms, enables Li-ion to be charged in the inconsistent manner previous only accepted by sealed lead acid.

6.3.2 Battery Charge Management

The charge method for Li-ion cells is a constant current charge [9], followed by a constant voltage charge phase. This charge is positively terminated when the current sourced by the battery pack drops below a predetermined maximum charge rate threshold. A maximum charge rate is specified by the cell manufacturer, and can be found on the cell data sheet. The minimum battery charge time is given, based on the maximum allowable charge rate. Once this voltage is reached, the charger will source only enough current to hold the voltage of the battery at this constant voltage. The accuracy of the set point voltage is critical. If this voltage is too high, the number of charge cycles the battery can complete is reduced, shortening the battery life. If the voltage is too low, the cell will not be fully charged.

Charging at very low and very high temperatures is damaging for Li-ion batteries. The allowable temperature window is quite small. In many applications the maximum allowable temperature is highly problematic for consistent reliable battery charging.

Charging methods for Nickel chemistries (NiCd and Nimh) are separated into two general categories: Fast-charge is typically a system that can recharge a battery in about 1 or 2 h, while slow charge usually refers to an overnight recharge.

Most high-performance charging systems employ at least two detection schemes to terminate fast-charge. Voltage or temperature is typically the primary method, with a timer for insurance if the primary method fails to correctly detect the full charge point.

The full charge point can be determined by sensing the cell temperature or cell voltage. Temperature sensing is preferable to voltage sensing because the cell temperature gives the most accurate information about what is happening within the cell. However, if the cell temperature is to be accurately measured, the temperature sensor must be built into the battery pack which increases the manufactured cost of the battery. Two temperature techniques include the delta T detector and the temperature slope detection.

In the temperature slope detection, a circuit which measures the rate-of-change or slope of the cell temperature can be used for end-of-charge detection with both NiCd and Nimh batteries. This type of circuit is referred to as a dT/dt detector, because it measures the change in battery temperature with respect to time. Temperature slope detection is typically used in processor-based systems. Temperature readings are taken at timed intervals, and stored in memory. The present temperature reading is compared to the previous value and the difference during that time period is calculated. Once the temperature change over a timed interval is known, the rate-of-change or slope of the cell temperature is calculated and compared to a target value. When the target is reached, the fast charge is terminated because the battery is fully charged.

However, voltage sensing is easier, because the voltage leads are easily accessible and require no special assembly in the battery pack. Voltage slope detection is common in processor-based systems that can measure, store, and compare battery voltage readings taken at timed intervals can accurately detect end-of-charge by using a method called voltage slope detection. This method of charge termination can be used with both NiCd and Nimh, as long as the system accuracy, resolution, and noise immunity are adequate for the job.

In general, nickel, and particularly NiCd, can accept higher charge rates or *C* rates than Li-ion and therefore can be charged faster.

For portable devices, knowing the state-of-charge (SOC) of a battery and its relation to the expected run time is critical for the user experience. The popular chemistries discussed rely on a process in the battery pack to provide this and other information to the host processor and the battery charger. There are accepted popular protocols for communicating with processor, or the smart battery packs. These include the system management bus (SMB), the two wire I^2C protocol that is part of the SBS, or smart battery system standard, and HDQ, or DQ, which is a single wire serial interface.

The advantages of software controlled charging is the unlimited feature. In a given application, typically only one extra feature is necessary to tip the scale in its favor.

The state-of-charge can easily and inexpensively be delivered to end-users with a four or five segment LED, or displayed on a LCD. Most communication protocols are available via the microprocessor. Battery management and maintenance features, sometimes critical for end-users, are available at minimal additional expense. These features include design versus relative capacity, how much energy the battery pack can hold as compared to when it was new.

In addition battery management capabilities include gas guage management for fixing the gas gauge that has lost its calibration. Battery conditioning for performing discharge and charge cycles to check the health of a battery. In addition reset gas gauge registers, notifies a user when a pack has had a preset number of cycles indicating the need to replace the battery.

PWM control by the processor nets an additional set of benefits, including varying the charge current to minimize temperature rise, when the ambient temperature is already high. This is particularly important when charging Li-ion batteries, with their narrow charge temperature window, and can mean the difference between charging and not charging. In addition, variable charge voltages, for different pack voltages, including voltages required for emerging chemistries and express or fast charging, where the charger can have adaptive charge currents based on available power, temperature conditions, and maximum allowable charge current of the battery.

Weaknesses of software controlled charging are few in number. However, they are not insignificant. The most important drawback is the development of the software. There are the time and costs to develop it, and verify it. In low power, single power implementations, there is typically a price premium to pay, versus a dedicated charge control IC. Lastly, part of that premium in both development time, and recurring costs, is a 5 V power supply needed for the microprocessor.

An external linear regulator will supply a nominal DC voltage for use by the mobile handset's built-in charger-and-safety circuit. In some cases, the Universal Serial Bus (USB) provides the DC input for battery recharging. This enables business travelers to shed the outboard linear power supply, and refresh their mobile handsets from the USB port on their personal computers. The vast majority of users, however, will continue to recharge their phones from an external power supply plugged into a wall outlet.

6.3.3 Li-Ion Battery Safety

It is critical for Li-ion battery pack manufacturers to build safe and reliable products for battery-powered systems. Battery management electronics in battery packs monitor Li-ion battery operating conditions, including battery impedance, temperature, cell voltages, charge and discharge current, and SOC to provide detailed remaining runtime, and battery health information to the system to ensure that right system decisions can be made. Additionally, to enhance battery safety, whenever at least one of the fault conditions, such as over-current, short circuit (SC), cell and pack over-voltage, or over temperature, the battery cells are disconnected from the system by turning off two back-to-back protection MOSFETs that are in series with the Li-ion cells in the battery pack. Impedance track technology-based battery management unit (BMU) monitors battery cell impedance over the entire battery life cycle as well as cell voltage imbalance, potentially capable of detecting the cell micro-short and preventing the cell from fire hazard or even explosion.

Excessive high-level operating temperatures accelerate cell degradation and causes thermal run-away and explosions in Li-ion batteries. This is a specific concern with this type of battery because of its highly aggressive active material. Rapid temperature increase can occur if a battery is overcharged at high current or shorted. During overcharge of a Li-ion battery, active metallic lithium is deposited on an anode. This material dramatically increases the danger of explosion, because it can explosively react with a variety of materials including electrolyte and cathode material. For example, Lithium-carbon intercalated compound reacts with water and the released hydrogen can be ignited by the heat of the reaction. Cathode material, such as $LiCoO_2$, starts reacting with electrolyte when the temperature exceeds its thermal run-away threshold of 175°C with 4.3 V cell voltage.

Li-ion cells use thin, micro-porous films such as polyolefin to electrically isolate the positive and negative electrodes as they provide excellent mechanical properties, chemical stability, and are of acceptable cost. The low melting point of polyolefin, ranging from 135°C to 165°C, makes it suitable to be used as a thermal fuse. As the temperature approaches the melting point of the polymer, porosity is lost. This is intentional so it will shutdown the cell because lithium ions can no longer flow between electrodes. In addition, there is a PTC (positive temperature coefficient) device and a safety vent to provide additional protection in the Li-ion cells. The case, commonly used as the negative terminal, is typically Ni-plated steel. When the case is sealed, it is possible for the metal particles to contaminate the interior of the cells. Over time, the particles can migrate into the separator, degrading the insulating barrier placed between the anode and cathode sides of the cell. That creates a micro-short between anode and cathode, allowing electrons to flow freely, ultimately failing the battery. Generally, this type of failure leads to little more than the battery powering down and ceasing to function properly. In rare instances the battery can overheat, melt, catch fire, or even explode. This was reported as the main root cause of some recent battery failures that resulted in mass recall by different manufacturers.

6.3.4 Battery Authentication

As portable products become more commonplace, techniques need to be put in place to ensure that batteries and peripherals used by those products meet the original manufacturer's safety requirements. The selection of a battery authentication depends on the security level needed and cost for the applications. The options for the manufacturer include [10]:

- *Identification number*: The host controller can request an identification number that is answered by a fixed response.

- *CRC algorithm*: The host processor sends a random challenge and reads the response that is an encoding of the challenge and a shared secret key through the CRC with a shared secret polynomial.

- *SHA-1 encryption*: The host processor sends a random challenge and reads the response that is an encoding of the challenge and shared secret key through the SHA-1 cryptographic primitive.

The simple ID authentication is cheapest, but is least secure. The challenge and response-based authentication techniques are more expensive. However, they have highest security and are good for the high-end portable applications.

6.3.5 Example of a BMU and Battery Protection

The BMU provides the intelligence of the system for advanced functions such as fuel gauging calculations on remaining cell capacity, protection circuitry, and thermal sensors used to monitor internal battery pack temperature. In addition, LEDs that indicate cell or battery pack status, and a serial data communications bus that communicates to the host device. For very large battery packs with more than four cells in series, it is necessary to have intelligent cell balancing. This can be done during the charge cycle with bleed resistors, or with more advanced active cell balancing on both charge and discharge. A relatively new feature, authentication, provides improved safety and protection against aftermarket batteries, which are often of inferior quality.

Cell material developments are ongoing to increase thermal run-away temperature. On the other hand, although the battery must pass stringent UL safety tests such as UL1642, it is always going to be a responsibility of the system designer to provide correct charging conditions and be well-prepared for possibility of multiple failures of electronic components. The system should not cause battery catastrophic failures due to over-voltage, over-current, SC, over-temperature conditions, and external discrete component failures. This means redundant protection should be implemented – having at least two independent protection circuits or mechanisms in the same battery pack. It is also desirable to have the electronics circuit detect battery internal micro-shorts to prevent battery failures.

Figure 6.13 shows the BMU block diagram in the battery pack, which consists of gas gauge IC, analog front end (AFE) circuit, and independent second-level safety protection circuit.

The gas gauge circuit is designed to accurately report available capacity of Li-ion batteries. Its unique algorithm allows for real-time tracking of battery capacity change, battery

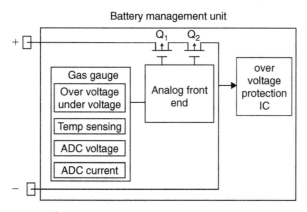

Figure 6.13: Battery Management Unit

impedance, voltage, current, temperature, and other critical information of the battery pack. The gas gauge automatically accounts for charge and discharge rate, self-discharge, and cell aging, resulting in excellent gas-gauging accuracy even when the battery ages.

A thermistor is used to monitor the Li-ion cell temperature for cell over-temperature protection, and for charge and discharge qualification. For example, the battery is usually not allowed to charge when the cell temperature is below 0°C or above 45°C, and is not allowed to discharge when the cell temperature is above 65°C. When over-voltage, over-current, or over-temperature conditions are detected, the gas gauge IC will command the AFE to turn off the charge and discharge MOSFETs Q_1 and Q_2. When cell under-voltage is detected, it will command the AFE to turn off the discharge MOSFET Q_2 while keeping the charge MOSFET on so that battery charging is allowed.

The main task of the AFE is overload, SC detection and protection of the charge and discharge MOSFETs cells, and any other inline components from excessive current conditions. The overload detection is used to detect excessive over-currents in the battery discharge direction, while the SC detection is used to detect excessive current in either the charge or discharge direction. The AFE threshold and delay time of overload and SC can be programmed through the gas gauge data flash settings. When an overload or SC is detected and a programmed delay time has expired, both charge and discharge MOSFETs Q_1 and Q_2 are turned off and the details of the condition are reported in the status register of AFE so that the gas gauge can read and investigate causes of the failure.

The AFE serves an important role for the gas gauge two-, three-, or four-cell lithium-ion battery pack gas gauge chipset solution. The AFE provides all the high voltage interface needs and hardware current protection features. It offers an I^2C compatible interface to allow the gas gauge to have access to the AFE registers and to configure the AFE's protection features. The AFE also integrates cell balancing control. In many situations, the SOC of the individual cells may differ from each other in a multi-cell battery pack, causing voltage difference between cells and cell imbalance. The AFE incorporates a bypass path for each cell. These bypass paths can be utilized to reduce the charging current into any cell and, thus, allow for an opportunity to balance SOC of the cells during charging. Since the impedance track gas gauges can determine the chemical SOC of each cell, a right decision can be made when cell balancing is needed.

The gas gauge has two tiers of charge/discharge over-current protection settings, and the AFE provides a third level of discharge over-current protection. In case of SC conditions when the MOSFETs and the battery can be damaged within seconds, the gas gauge chipset entirely depends on the AFE to autonomously shut off the MOSFETs before such damage occurs.

While the gas gauge IC and its associated AFE provide over-voltage protection, the sampled nature of the voltage monitoring limits the response time of this protection system. Most applications will require a fast-response, real-time, independent over-voltage monitor that operates in conjunction with the gas gauge and the AFE. It monitors individual cell voltages independent of the gas gauge and AFE, and provides a logic-level output which toggles if any of the cells reaches a hard-coded over-voltage limit.

The response time of the over-voltage protection is determined by the value of an external delay capacitor. In a typical application, the output of the second-level protector would trigger a chemical fuse or other fail-safe protection device to permanently disconnect the Li-ion cell from the system.

6.3.5.1 Battery Pack Permanent Failure Protection

It is critical for the BMU to provide conservative means of shutting down the battery pack under abnormal conditions. Permanent failure detection includes safety over-current discharge and charge fault, safety over temperature in discharge and in charge, safety over-voltage fault (pack voltage), cell imbalance fault, and shorted discharge FET fault and charge MOSFET fault. It is the manufacturer's choice to enable any combination of the above permanent failure detections. When any one of these enabled faults is detected, it will blow the chemical fuse to permanently disable the battery pack. As an extra fail-proof of electronics component failure, the BMU is designed to detect if the charge and discharge MOSFETs Q_1 and Q_2 failed. If either the charge or discharge MOSFETs is shorted, then the chemical fuse will also be blown.

Battery internal micro-short was reported as the main root cause for several recent battery recalls. How can you detect the battery internal micro-short and prevent the battery from catching fire and even exploding? The battery may have an internal micro-short when the metal micro-particles and other impurities from the case sealing process contaminate the interior of the cells. The internal micro-short significantly increases the self-discharge rate which results in lower open circuit voltage than that of the normal cells. Impedance track gas gauge monitors the open circuit voltage and, therefore, detects cell imbalance when the open circuit voltage difference between cells exceeds the preset threshold. When this type of failure happens, a permanent failure is signaled and MOSFETs are opened, and the chemical fuse can be configured to blow as well. This will render the pack unusable as a power source and thus screen the pack with the internal micro-short cells, thus, preventing it from causing hazards.

The BMU is crucial for the end-user's safety. The robust multi-level protections of over-voltage, over-current, over temperature, cell imbalance, and MOSFET failure detection significantly improve the battery pack safety. Impedance track technology can detect a battery internal micro-short by monitoring the cell open circuit voltage and disable the battery permanently, making the end-users safer.

6.4 PMICs Plus Audio

Next generation PMICs go beyond simple signal conditioning and distribution of power. They further ease system development by integrating a wide range of functions tailored to specific applications, such as full-featured audio paths with analog, digital, and power audio interfaces, touch screen support, coin cell backup supply switching and charging, backlighting, fun light LED drivers, and regulators optimized for specific functions such as RF.

A novel PMIC function permits users to interface portable devices directly to a car stereo through a USB cable or plug-in cradle. For improved utility in automotive applications, the emerging CEA-936 car kit standard is driving new capabilities for battery charging, data communications, and audio routing through a standard mini-USB connector. Suitably equipped products will allow hands-free phone calls or stereo playback by directly interfacing to a car kit-capable stereo. A USB transceiver with charging capabilities can be added to portable devices for under a dollar as a standalone IC or for even less as an integrated feature of a PMIC.

Integration of these functions along with their associated discrete devices and other low integration ICs not only improves board area and system cost but also allows coordinated control by a logic core with a direct link to the system processor and software. In this way, highly integrated PMICs can make the task of orchestrating the various system-level components much simpler since a multitude of functions can be accessed and controlled through a single programmable interface. System hardware design as well as software integration are streamlined through such functional consolidation.

The proliferation of voltage regulators are being curtailed by custom PMICs. These devices have as many as 20 separate voltage regulators on one chip. A combination of switch-mode regulators and LDOs, a battery monitor-and-charger, and perhaps a microprocessor supervisor and power rail sequencer all on one chip.

A number of PMIC devices incorporate audio playback codecs on the same chip with the power management functions. The Freescale MC13783 [11], for example, includes four buck regulators for the handset's processor cores, a boost regulator for backlighting, a wireless USB port, several LDO regulators, a Li-ion battery charger, a 10-bit analog-to-digital converter (ADC) for battery monitoring, a 13-bit voice codec, and a 16-bit stereo digital-to-analog converter (DAC) (supporting multiple sample rates). The stereo codec anticipates the inevitable transformation of the mobile handset into a portable media center with high-quality audio playback.

A high-level block diagram of the MC13783 is presented 6.14 indicating the wide functionality of the MC13783 including:

- 10-bit ADC for battery monitoring and other readout functions

- Buck switchers for direct supply of the processor cores

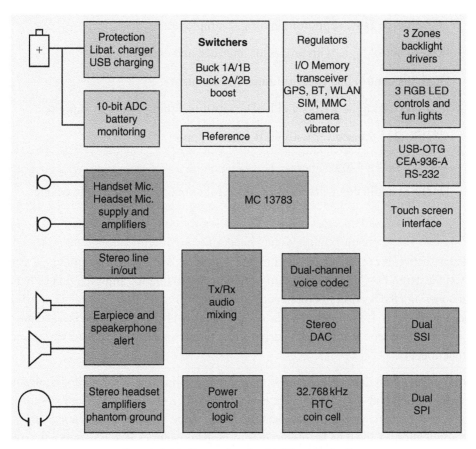

Figure 6.14: Example of a PMIC with Audio

Source: http://www.freescale.com

- Boost switcher for backlights and USB on the go supply

- Regulators with internal and external pass devices

- Transmit amplifiers for two handset microphones and a headset microphone

- Receive amplifiers for ear piece, loudspeaker, headset, and line out

- Battery charger interface for wall charging and USB charging

- 13-bit voice codec with dual ADC channel and both narrow and wide band sampling

- 13-bit stereo recording from an analog input source such as FM radio

- 16-bit stereo DAC supporting multiple sample rates

- Power control logic with processor interface and event detection

- Dual SPI control bus with arbitration mechanism

- Multiple backlight drivers and LED control including fun light support

- USB/RS232 transceiver with USB car kit support

- Touch screen interface

- Dual SSI audio bus with network mode for connection to multiple devices.

6.4.1 Audio

The audio section is composed of microphone amplifiers and speaker amplifiers, a voice codec and a stereo DAC. Microphone amplifiers are available for amplification of handset microphones and the headset microphone. Several speaker amplifiers are provided. A bridged ear piece amplifier is available to drive an ear piece. Also a battery supplied bridged amplifier with thermal protection is included to drive a low ohmic speaker for speakerphone and alert functionality.

A voice codec with a dual path ADC is implemented following GSM audio requirements. Both narrow band and wide band voice is supported. The dual path ADC allows for conversion of two microphone signal sources at the same time for noise cancellation or stereo applications as well as for stereo recording from sources like FM radio. A 16-bit stereo DAC is available which supports multi-clock modes. An on board PLL ensures proper clock generation. The voice codec and the stereo DAC can be operated at the same time via two interchangeable buses supporting master and slave mode, network mode, as well as the different protocols like I^2S.

6.4.2 Linear and Switching Regulators

Four buck converters and a boost converter are included. The buck converters provide the supply to the processors and to other low-voltage circuits such as I/O and memory. The four buck converters can be combined into two higher power converters. Dynamic voltage scaling (DVS) is provided on each of the buck converters. This allows, under close processor control, the ability to adapt the output voltage of the converters to minimize processor current drain.

The boost converter supplies the white backlight LEDs and the regulators for the USB transceiver. The boost converter output has a backlight headroom tracking option to reduce overall power consumption. The regulators are directly supplied from the battery or from the switchers and include supplies for IO and peripherals, audio, camera, multimedia cards, SIM cards, memory, and the transceivers. Enables for external discrete regulators are included as well as a vibrator motor regulator. A dedicated preamplifier audio output is available for multi-function vibrating transducers.

Drivers for power gating with external NMOS transistors are provided including a fully integrated charge pump. This will allow to power down parts of the processor to reduce leakage current.

6.4.3 Battery Management

The MC13783 supports different charging and supply schemes including single path and serial path charging. In single path charging the phone is always supplied from the battery and therefore always has to be present and valid. In a serial path charging scheme the phone can operate directly from the charger while the battery is removed or deeply discharged.

The charger interface provides a linear operation via an integrated DAC and an unregulated operation used for pulsed charging. It incorporates a standalone trickle charge mode, in the case of a dead battery, with a LED indicator driver. Over-voltage, SC, and under-voltage detectors are included as well as charger detection and removal. The charger includes the necessary circuitry to allow for USB charging and for reverse supply voltage to an external accessory. The battery management is completed by a battery presence detector and an A to D converter that serves to measure the charge current, battery, and other supply voltages as well as measuring the battery thermistor and die temperature.

6.5 Summary

The proliferation of power management functions in the typical mobile phone, and the volumes associated with each phone model, argues for customization and ASIC integration. While the need to manipulate voltages and currents are contrary to the use of fine-geometry CMOS and integration with the digital baseband or applications processor, PMICs with many voltage regulators on one chip can be constructed with bipolar CMOS and CMOS technologies.

A new standalone PMIC includes 20 separate voltage regulators, a boost converter for white LED drives, a built-in battery monitor-and-charger, and a USB interface.

Device types and functions employed in portable devices are shown in Table 6.2.

From a regulator perspective, PMICs generally offer one to three types of programmable regulators – LDO, buck, and boost – which convert battery voltage to a stable voltage for a particular component. LDOs with integrated pass FETs are cost-effective, simple to use, and only need a filtering capacitor on the output. However, these conveniences come at the expense of efficiency. For example, a 200 mA LDO with a 3.6 V input supply and a 1.2 V output will run at about 33% efficiency, with the losses dissipated as heat in the pass device.

In contrast, a buck switching regulator operating under the same conditions may run at about 90% efficiency since the power paths flow through transistors operating in the low resistance triode region, minimizing the $I*R$ product and hence reducing wasted power. However, buck and boost switching regulators take up more silicon area than LDOs and also require an inductor on the output, which increases overall system cost. A PMIC may provide as many as 10 or more LDOs and 1 or more each of bucks and boost regulators. A buck switcher is often utilized for the main power consumer in a system, most likely an application or baseband processor, where high efficiency will yield the most benefit. In this way, designers can optimize power efficiency with cost.

For advanced systems such as 3G WCDMA phones, PMICs offer multiple buck regulators to support independent high-efficiency zoning of multiple processors (core, baseband, application, etc.), memory, and/or high power peripherals. Programmable buck regulators are critical to further optimize power consumption through DVS, an increasingly important feature in advanced processors, which dynamically adjust their clock frequency and voltage, to reduce power consumption when the processor is not 100% utilized.

Smart power devices will continue to drive advances in power efficiency and enable new interface capabilities in portable products while extending battery life as new features

Table 6.2: Power Management Devices, Functions, and Vendors

Devices	Functions	Vendors
Buck-boost switching regulators	Low-voltage switching regulator that steps down or steps up battery voltage outputs	Linear Technology
LDOs	Linear regulator most efficiency in low-voltage, low-current applications	National Semiconductor, Linear Technology, and Micrel
Switch-mode RF power amplifier drivers	Switching regulator whose output waveform approximates mobile phone callers' voice	Maxim and National Semiconductor
White LED backlight drivers	Step-up switching regulator providing constant current for LED arrays	Maxim, Linear Technology, and Analogic Tech
Power management ICs (CPU monitors)	Monitors CPU instruction cycles to induce "sleep,""standby," or "wakeup" modes	National Semiconductor
Battery management ICs	Fuel gauges and charging controllers for rechargeable batteries	Texas Instruments and Intersil
Integrated power management devices (PMICs)	Custom circuits with as many as 10 voltage regulators on 1 chip	Texas Instruments, Dialog Semiconductor, and Qualcomm

are added and product performances are enhanced. By integrating power management and user interface functions within a single chip, PMICs can facilitate control of power consumption for optimization at the system level, enabling designers to easily and efficiently maximize battery life.

References

[1] Selecting Power Management ICs for Cellular Handsets, http://www.maxim.com. Application Note APP 3174, April 06, 2004.

[2] S. Davis. LDO Voltage Regulators, http://www.nsc.com, 2006.

[3] M. Day. Understanding Low Drop Out (LDO) Regulators, http://www.ti.com, October 2006.

[4] L. Zhao and J. Qian. DC–DC Power Conversions and System Design Considerations for Battery Operated System, http://www.ti.com, October 2006.

[5] O. Nachbaur. Evaluation and Performance Optimization of Fully Integrated DC/DC Converters, http://www.ti.com, October 2006.

[6] F. Franc. Multiple Linear Regulators Power Handheld Devices, http://www.commsdesign.com, July 2002.

[7] Battery Management, http://www.ti.com, October 2006.

[8] Gas Guaging Basics Using TIs Battery Monitor ICs, http://www.ti.com, October 2003.

[9] J. Formenti. Optimization of the Charger to System Interaction Enables Increased Handheld Equipment Functionality, http://www.ti.com, October 2006.

[10] T. Vanyo and J. Qian. Battery Authentication Architecture and Implementation for Portable Devices, http://www.ti.com, 2006.

[11] N. Cravotta. Longer Battery Life through Integrated Power Management, http://www.edn.com, November 2004.

CHAPTER 7

System-Level Approach to Energy Conservation

7.1 Introduction

The challenging energy demands of new portable devices are forcing radical approaches to solve complex power management problems. It means expanding the definition of power management beyond power delivery, to include power distribution and power consumption. It means interfacing the power-delivery system to the power consumption system, to permit the system to communicate with each other to vastly improve power conservation.

To achieve these quantum leaps manufacturers need a new industry model. Historically power management and processor integrated circuit (IC) suppliers have developed their technologies independently. However, power efficiencies are now reaching levels where only minimal gains can be achieved through this conventional-isolated approach. No longer are piecemeal short-term solutions addressing the power efficiency of individual components. Rather the entire system has to be considered as a whole and the opportunities for system components to work together have to be leveraged to obtain power performance level required by next generation devices.

Effective power management requires more than simply turning devices on or off. A flexible power architecture is needed. For example, many processors and peripherals support multiple power modes such as Standby and Sleep to power down unused functions except when they are in active use. Multiple power zones are required to support the various supply requirements and specialized operational modes to extend battery life for portable products. Each zone is supplied by an independent linear or switching regulator, giving developers a high degree of control over power.

Further optimization can be achieved through techniques such as coordinated wake-up sequences; instead of turning components on all at once, power is conserved by firing up various circuits only when they are needed. For example, in a cell phone, there is no point in powering analog amplifiers before the radio frequency (RF) synthesizer has stabilized, nor turning on the digital signal processor (DSP) before a signal has been received that needs processing. Using tuned and coordinated wake-up sequences helps to manage the supply perturbations due to current surges flowing through the source impedance of the battery. By spreading out startup surges and reducing battery dips, the end-of-life cutoff for minimum battery level can be extended with a resulting improvement in effective battery lifetime between charge cycles.

With a highly integrated approach, a power management integrated circuits (PMICs) enables developers to optimize power consumption at the system level, for specific applications, while significantly reducing the design complexity required to achieve these gains. For this reason, early consideration of power management details is fundamental for designing power efficient hardware and software that can be relied upon to provide optimal battery life.

Solutions require a systems approach where portions of the power management circuitry exist in separate ICs and portions are integrated into the system chips. Companies such as AMD, Intel, Texas Instruments, Freescale, ARM, National Semiconductors, and Transmeta have obtained good results using this type of frequency–voltage management in addition to clock gating.

7.2 Low Power System Framework

Basic Energy Management System: Energy Management System (EMS) employs techniques to minimize the drain from the battery or other limited power source. The diagram shows a basic EMS solution comprising three parts:

1. The Platform Hardware with various power managed components (PMC).

2. Performance Estimator employing dynamic voltage and frequency scaling (DVFS) and other power saving algorithms.

3. Performance Setter that maps the estimations to specific platform operating points (OP), i.e. combinations of PMC settings.

The best operating settings for the PMCs depends on the changing workload demanded by the processor, peripherals, and the software. This is application specific, and typically several application programs run concurrently.

One approach is to characterize the workload for each application before hand and cumulate for all known use cases. However, this is sub-optimal and an efficient solution would use prediction and other techniques to compute the required performance in real time (Figure 7.1).

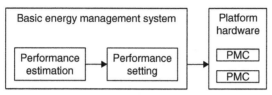

Figure 7.1: Basic Energy Management System

7.2.1 Advanced Energy Management Solution

The EMS dynamically adjusts the performance-power settings of the device hardware to levels that achieve the appropriate level of performance to handle the software workload within its real-time deadlines.

The Performance Estimator uses additional performance predictors and system event monitors to improve its performance estimation. Incrementally, it could have an interface to "power-aware" application programs and middleware software capable of specifying their required performance levels dynamically.

The Performance Estimator computes the actual workload from moment to moment. However, the performance setting-function does not change the hardware settings for each new estimation due to the cost, measured in time and power, in changing from one operating point to another. An advanced EMS performs a cost–benefit analysis in real time to decide whether a change in OP makes the energy saving better or worse. For example, processor low power idle modes, like Sleep and Doze, each have a "break-even" time. Unless the device spends at least the break-even time in the low power mode, you would be better of, in energy-saving terms, staying in the Run mode. If the device has reliable event-driven information to determine how long it can be idle for, the EMS can make intelligent selections of which low power mode it can employ as required (Figure 7.2).

An advanced EMS systematically encodes hardware design data so that its run-time choices are optimized for the chipset on which it is running. These are called System Cost Rules.

7.2.2 Software for Self-Optimizing Systems

The EMS algorithms are fed with relevant real-time events to make minor corrections for changing scenarios. These minor corrections include adapting the optimal device settings in response to product user inputs via user policies.

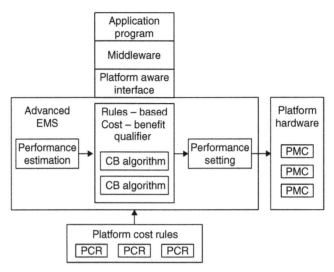

Figure 7.2: Advanced Energy Management System

A "user-friendly" advanced EMS should allow the device developer to set approximately optimal settings in the EMS for a broad set of software use cases and applications. The adaptive nature of the EMS manages the run-time variations. This is not just software optimization but software for self-optimizing systems. The concepts developed in EMS are the foundation of a number of available power conservation techniques.

7.3 Low Power System/Software Techniques

Power-saving technologies have been developed to address sources of power waste. Many are hardware solutions such as smaller transistor geometries and active well biasing. However, other hardware technologies require software and different software techniques must be used to exploit each of these different hardware technologies effectively. Dynamic power management (DPM) describes a system that sets the power states of its hardware modules in real time to minimize power waste, and still meets performance needs. DPM includes techniques such as dynamic voltage and frequency scaling (DVFS), dynamic process and temperature compensation (DPTC), and idle time prediction for controlling low power idle modes (Doze, Sleep, Hibernate, etc.)

The software techniques shown in Figure 7.3 include Dynamic Voltage Frequency Scaling and Dynamic Process Temperature Compensation. Application programs are monitored during execution by one or more of the software techniques illustrated. Some of these applications know their performance-power needs ("power-aware" software) and others do

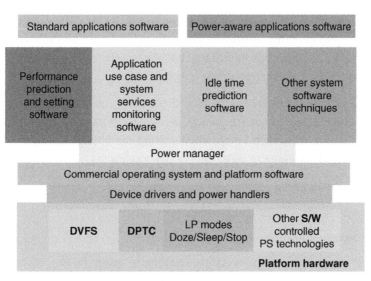

Figure 7.3: Power-Saving Software Techniques

Source: www.freescale.com

not. These techniques are employed to control a power manager that drives the hardware power-saving mechanisms using software drivers and power handlers in the operating system.

7.3.1 Dynamic Frequency Scaling

System-on-chip (SoC) dynamic power consumption is proportional to operating frequency. It makes sense to lower the clock frequency of a processor to the lowest value that still meets the required processing performance. Although the software runs more slowly, it still meets its real-time deadlines with acceptable margins. This is performed dynamically and requires special power management software to compute which frequency setting is acceptable.

It may seem intuitive that although this lowers the instantaneous power consumption, it might not reduce the overall energy requirement, but instead spreads it over a longer period. In fact, some memory systems, such as integrated level 2 caches and SDRAMs, will incur fewer accesses at the reduced operation frequency, so their circuits incur reduced switching and therefore reduced power consumption. However, the benefits of frequency scaling alone to total energy management are marginal.

7.3.2 Dynamic Voltage Scaling

Since power varies with the square of voltage, square-law power savings are possible with voltage scaling. If voltage scaling and frequency scaling are both used, the combination, called DVFS, yields power savings roughly proportional to the cube of operating voltage. These square-law and cube-law power savings depend not only on the configuration and efficiency of the voltage control circuits, but also on the efficiency of the prediction software used to set the voltage/frequency settings.

For a given SoC design, the operating voltage determines the maximum usable operating frequency. The voltage, and hence frequency, are scaled to trade required performance against minimal power waste. However, when scaling the voltage the operating frequency must be scaled with the voltage to meet the design constraints.

Figure 7.4 shows the required sequence for changing:

- First, to a higher voltage and frequency, and then
- Later, back to a lower voltage and frequency.

At point 1 a request is made to the voltage controller to ramp the processor voltage up to the new higher value. The voltage will increase at a defined slew rate, to avoid excessive power surges, and the controller will notify the requester, via a processor interrupt, when the target voltage has been reached at point 2. The processor must continue at the lower frequency until then, at which point software then signals its clock generator, typically

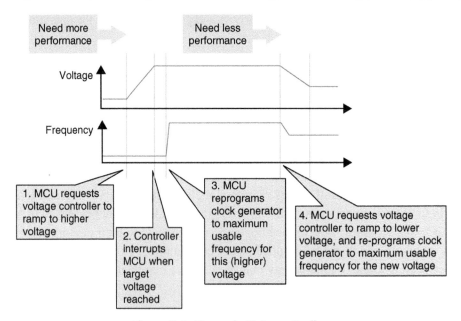

Figure 7.4: Dynamic Voltage Scaling

Source: www.freescale.com

a programmable Phase Lock Loop, to change the processor frequency to a higher one, usually the highest usable frequency for the new voltage, which occurs at point 3. The frequency change is quick compared to the slow slew rate that must be adopted.

When going to lower voltages/frequencies, the processor changes its frequency and requests the voltage change at point 4. The processor need not wait until the voltage ramp down has completed before it continues normal software execution, since the new lower frequency is already within the operating range of the voltage profile.

7.3.3 Dynamic Process and Temperature Compensation

The operating voltage setting chosen with DVFS to achieve a specific operating frequency is a worst-case value. It includes a voltage margin for variations in process and temperature. This margin represents power waste because the SoC is operating at a slightly higher voltage than it needs for the operating frequency. By monitoring the SoC using an on-chip process and temperature-dependent structure, it is possible to calculate a lower operating voltage that is very close to process limits, which thereby minimizes power waste. If the SoC is manufactured in two or more silicon processes that information could also be fed into the calculations (refer back to Chapter 3).

7.3.4 Handling Idle Modes

DVFS technology addresses varying but continuous software workloads. However, what happens when the processor or other power-hungry device has periods of no activity? When the processor has periods of inactivity, the hardware devices have low power idle modes that software can exploit to save power. However, some devices have more than one idle mode. A processor core could have Doze, Sleep, and Stop modes. The deeper modes yield more power savings, but usually come with a greater "cost" penalty in time and transition power when entering and exiting the mode. For example, an idle mode based on power gating requires the hardware's internal memory state to be saved and restored, either by hardware or software.

In a real-time system, the device's run state must be re-entered in time-to-service events that otherwise would degrade the user's quality of service. Smart software must either predict or have advanced knowledge of an event and the time taken to restore the hardware from idle to run mode. Based on the predicted start and end of an idle period, it controls the shut-down and wake-up mechanisms associated with the idle modes. The prediction software has to determine which of the idle modes will yield a net power saving for each idle period it encounters.

7.4 Software Techniques and Intelligent Algorithms

A simple DVFS software driver attached to the operating system is used to interface to the integrated DVFS control logic to increase or decrease the processor's operating frequency and voltage. Controlling and monitoring the DVFS mechanism is the easy part.

Knowing in real time which frequency/voltage setting to use, and when, is more challenging. Intelligent software is needed to compute dynamically how much processor performance the application programs and other system software require at any given time.

How intelligent that software is will determine how close you can get to the theoretical hardware limit on power saving. Higher performance software is more complex and more costly to develop. There is a performance versus complexity/cost trade-off which will vary for different applications and products. The trick is to develop intelligent energy management technology that is both flexible and scalable to give improved performance at acceptable cost across a wide range of devices. That includes the ability to support one or more algorithms for many hardware power-saving technologies.

7.4.1 Operating System

Typically, each microprocessor and DSP in wireless applications uses operating system software to manage the large number of hardware and software resources. A DPM approach should treat the OS as a state machine in which each state may require its own power management techniques. For example, DVFS could be used using normal software execution of tasks and Operating System, idle mode predictors used during software idling, and full-speed execution during interrupt handling.

Figure 7.5 shows a possible architecture for the power management software for a modern applications domain operating system. The hardware modules may each have several power states including a fully powered, and one or more low power idle states. A DVFS-controlled processor will have multiple active states and may also have several idle states.

The instantaneous values of the collective power states could be related to the phone use modes in many ways. The main criterion for judging an algorithm is the average power saving obtained over a range of typical use cases compared to a system which has no software power minimization.

Performance-prediction and performance-setting algorithms are used to control the performance-power states of the system hardware dynamically. These algorithms are very sophisticated and may need to be adapted or tuned to suit different system designs.

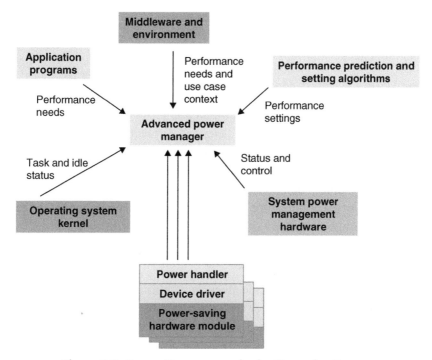

Figure 7.5: Power Management in the Operating System

Source: www.freescale.com

7.4.2 Typical DVFS Algorithm

Many algorithms exist for use with DVFS-based processors that set the processor's operating frequency and voltage based on predicting the short-term software workload on the processor.

An example algorithm in this class tracks the recent software workload history of each task running in the OS. This technique assumes a reasonable correlation between the recent past workload of a task and that of the near future. The task status information must be supplied by the OS kernel.

The algorithm maintains estimates of workload and unused idling time to predict the aggregate workload for all tasks. This normalized MCU processing level is translated by associated software into the relevant frequency and voltage settings required for the specific DVFS mechanism used. The algorithm continuously re-calculates and supplies new predictions in response to changing software workloads.

Ideally, the algorithm correctly predicts the required processor performance that meets individual deadlines for each OS task. The algorithm works well for OS tasks whose workloads do not change very rapidly.

7.4.3 Scope Within Wireless Mobile Applications

Some may question whether such complex prediction techniques are really necessary in a mobile application. After all, most of the "engine" software that runs on the cellular modem DSP, the Bluetooth® and WLAN processors, although it involves complex protocols, is fairly deterministic and can be well characterized and optimized for energy conservation at design time. The main and largely unexploited opportunity for large power savings exists in the applications domain. In the application domain the processor and multimedia applications, such as video and audio, run in real-time.

7.5 Freescale's XEC: Technology-Specific Intelligent Algorithms

For power-managed platforms, Freescale takes an architectural approach, and creates a generic software/hardware XEC framework with applications programming interfaces (APIs) and hardware interfaces. Its purpose is to help to maximize the portability of XEC technology across Freescale platforms.

Prediction-based software control of power-saving techniques is an immature technology and is just starting to emerge within the mobile wireless device industry. As yet there are no existing, widely accepted power management standards among mobile device manufacturers, operating system vendors, semiconductor companies, and others.

Algorithmic software, known as performance predictors, is used to predict runtime workloads of specific modules or power-saving technologies (e.g. DVFS). A predictor module can be included or removed from the software build depending on whether or not its corresponding hardware technology is present.

An advanced energy-saving solution such as XEC uses two or more performance predictors for certain technologies such as DVFS, to ensure high performance under a wide range of operating conditions (Figure 7.6).

Typically, a performance predictor operates abstractly. For example, a DVFS predictor only tries to predict required processing power as a normalized fraction of the maximum possible. It does not deal with the processor frequencies, operating voltages, MIPS, etc.

These device specific details of the DVFS hardware technology are handled by the XEC framework and OS driver software. Ideally, performance predictors are designed to be

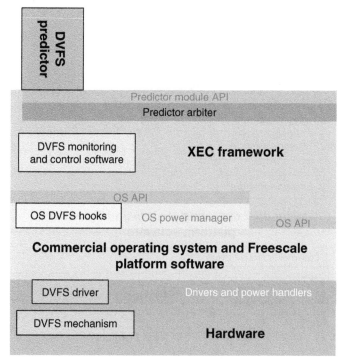

Figure 7.6: DVFS Predictor

Source: www.freescale.com

independent of the specific OS and hardware, to enable easy porting from one platform to another.

Depending on the software activity, the processor and other hardware modules will have periods of idling where they could be put into low power idle modes. The XEC technology uses an idle time predictor. It contains prediction software with connections to the OS and other event-monitoring software. The monitoring and control software specific to the platform's idle mode hardware is located in the XEC framework and in the OS device drivers for the hardware (Figure 7.7).

In principle, any hardware module with software-settable power-saving modes can be managed by the idle time policy software for minimum power waste (Figure 7.8).

7.5.1 XEC Framework

The XEC framework isolates the algorithmic components, namely the performance predictors, from the details of the specific platform OS and hardware. It contains

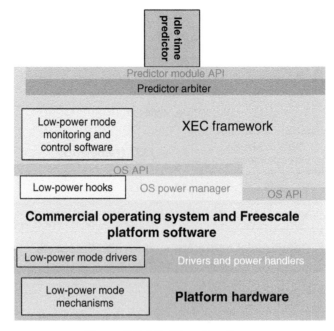

Figure 7.7: Idle Time Predictor

Source: www.freescale.com

platform-specific mechanisms for controlling and monitoring hardware power-saving technologies such as DVFS. In addition, it also provides common APIs for power-aware application programs and replaceable XEC modules like the performance predictors.

Because performance predictors may operate concurrently, a predictor arbiter arbitrates between multiple predictors to resolve which of their performance-power recommendations to select at any given time. The arbiter is programmable, so that the priority of any one performance predictor may be changed during normal operation depending on the runtime circumstances. Another framework component called the policy manager, acting as an agent for the product user, dynamically selects a policy, from a group of policies that sets the criteria for trading off performance, power, and energy in general or for specific situations or programs.

The XEC framework uses platform-specific power-cost rules to determine when and if to transition a particular hardware module from its current power state. Finally, the framework has an OS adaptation layer containing OS-specific code. Its purpose is to minimize the porting effort and configuration management problems to migrate the XEC software to the operating system.

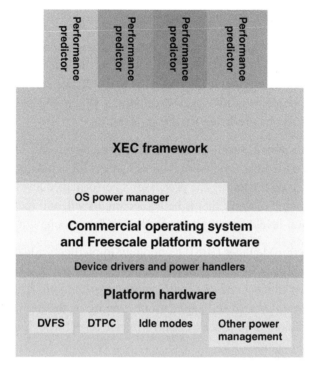

Figure 7.8: Performance Predictors

Source: www.freescale.com

Based on advanced runtime performance algorithms, XEC uses a standardized software framework that supports multiple concurrent predictors, policies, power-cost rules, etc. It runs as system software with commercial operating systems. The XEC solution targets more potential areas for energy savings such as SoC-specific low power modes and LCD panel hardware.

XEC is transparent to application programs and middleware, unless that software is optimized as a "power-aware" application; in that case, the application could communicate its power needs to XEC.

Using advanced techniques, XEC dynamically discovers each piece of software's required performance, sets hardware power management features like DVFS, multiple low power idle modes, etc., for just-enough performance and minimal power waste, and achieves much more energy savings than traditional techniques. XEC has performance predictors, real-time cost–benefit analyzers, policies and a policy manager, and many other advanced algorithms.

7.6 ARM's Intelligent Energy Manager

Completing a task before its deadline, and then idling, is significantly less energy efficient than running the task more slowly so that the deadline is met exactly. The goal of ARM's Intelligent Energy Manager (IEM) is to reduce the performance level of the processor without allowing applications to miss their deadlines. The central issue is how the right level of performance can be predicted for the application.

ARM has developed a system level solution to energy management. The IEM [1] framework provides a hardware and software mechanism for achieving these goals: it standardizes the interface for setting the processor's performance level, specifies counters for measuring the amount of work that is being accomplished, and includes operating system and application-level algorithms for predicting future behavior.

ARM Intelligent Energy Manager (IEM) technology implements advanced algorithms to optimally balance processor workload and energy consumption, while maximizing system responsiveness to meet end-user performance expectations. The IEM technology works with the operating system and applications running on the mobile device to dynamically adjust the required processor performance level through a standard programmer's model (Figure 7.9).

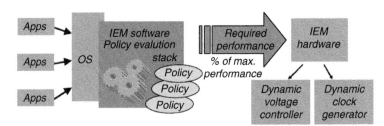

Figure 7.9: IEM Technology Controlling Voltage and Frequency [2]

The IEM software component uses information from the OS to build up a historical view of the execution of the application software running on the system. A number of different software algorithms are applied to classify the types of activity and to analyze their processor utilization patterns. The results of each analysis are combined to make a global prediction about the future performance requirement for the system. This performance setting is communicated to the hardware component so that the SoC-specific scaling hardware can be controlled to bring the processor to that level of performance.

7.6.1 IEM Policies and Operating System Events

IEM Policies are short, modifiable software routines that control what to do when an OS event occurs. An IEM kernel employs policies to determine a response to an event like increase or decrease the voltage and frequency to the CPU. Some software tasks are predictable in their behavior and can track past behavior to predict future performance requirements. IEM policies and their relationship to the IEM software architechture are illustrated in Figure 7.10.

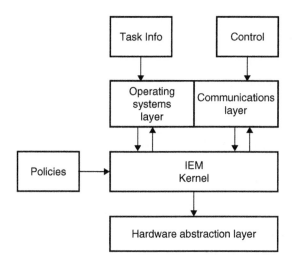

Figure 7.10: IEM Software Architecture

An event occurs in the operating system. For example, new task starts or battery low detected. The operating system informs IEM and policies checked to determine the appropriate action. A new frequency requirement is determined and sent to the performance controller.

7.6.2 Types of policy

Figure 7.11 indicates a conceptual view of IEM policies [3].

A *Step Policy* monitors the time spent by the operating system in idle task. If the time is large it reduces the frequency and if small it increases the frequency.

A *Perspective Policy* monitors the performance from the perspective of the task. The policy minimizes idle time available between each time the task is run. In addition it removes the effect of the task being pre-empted.

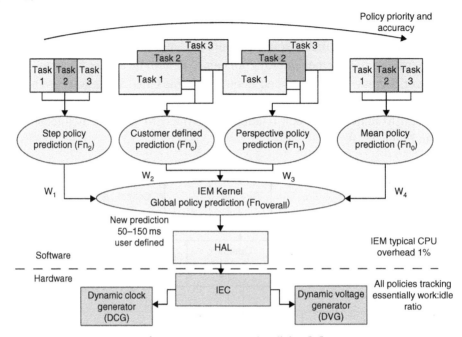

Figure 7.11: Types of Policies [3]

A *Mean Policy* generates the average workload for tasks recently run. If more time is available, the policy reduces frequency the next time the task is run. If the task did not complete in time the policy increases the frequency the next time.

Customer Policies are created by customers for applications such as gaming or playing music or events like low battery.

Polices are split into standard part and fast part. The standard part is run at a scheduled interval and may be pre-empted by higher priority tasks. However, the fast part is run during every OS event including context switch, task creating and deletion. It is preemptible only if the OS supports pre-emption of the kernel.

The fast part of the policy is used to capture time-critical information that will not be available when the standard part runs. In addition it reacts quickly to events like real time applications being scheduled to run and to capture information that is not accessible to the standard part.

The IEM software layer has the ability to combine the results of multiple algorithms and arrive at a single global decision. The policy stack supports multiple independent performance-setting policies in a unified manner. The primary reason for having multiple policies is to allow the specialization of performance-setting algorithms to specific situations, instead of having to make a single algorithm perform well under all conditions. The policy stack keeps track of commands and performance-level requests from each policy and uses this information to combine them into a single global performance-level decision when needed.

Using this system, performance requests can be submitted any time and a new result computed without explicitly having to invoke all the performance-setting policies. While policies can be triggered by any event in the system and they may submit a new performance request at any time, there are sets of common events of interest to all. On these events, instead of re-computing the global performance level each time a policy modifies its request, the performance level is computed only once after all interested policies' event handlers have been invoked.

7.6.3 Generic IEM Solution

The IEM system components, illustrated in Figure 7.12, [4] include:

- *Intelligent Energy Manager (IEM)*: A software component that measures the workload and predicts performance level needed to achieve it.

- *Intelligent Energy Controller (IEC)*: A hardware block that assists the measurement of the workload and transforms requirement to a percentage frequency value.

- *Dynamic Voltage Controller (DVC)*: A hardware module that transforms desired performance level to voltage and frequency via an open and closed-loop approach. Adaptive Power Controller (APC) from National Semiconductors is an example of a DVC.

- *Power Supply Unit (PSU)*. A highly controllable and responsive power supply.

The IEM software and hardware monitor the system workload to generate a performance request. The DVC can then set the correct operating voltage in either open-loop or closed-loop mode without processor intervention. The DVC will transparently provide the fastest possible response while assuring that the processor will always receive the minimum safe operating voltage for any given clock frequency. The DVC would also coordinate all clocks switching including the verification of stable supply voltage. The IEM provides a uniform software interface to simplify implementation and reuse. The DVC provides an open-standard interface to the external power supply.

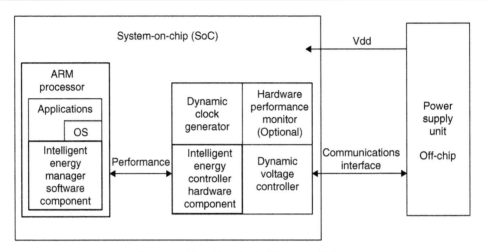

Figure 7.12: IEM Block Diagram [4]

7.6.4 Intelligent Energy Controller

The IEC is an advanced microcontroller bus architecture (AMBA) bus-compliant peripheral which provides a number of support functions to the IEM software. The IEC provides an abstracted view of the SoC-specific performance scaling hardware. It is responsible for translating the performance prediction made by the IEM software (0–100% of maximum performance) into an appropriate performance point at which the system will run and then controlling the scaling hardware to achieve operation at that point.

The IEC also measures the work done in the system to ensure that the software deadlines are not going to be missed. Additionally, the IEC supports a "maximum performance" hardware request feature.

The three major parts of the IEC, illustrated in Figure 7.13, are:

1. An AMBA Advanced Peripheral Bus (APB) interface which provides the IEM software with a standard API interface and allows it to communicate with the dynamic performance controller (DPC) and dynamic performance monitor (DPM) blocks. It also provides the IEM software status information about the support provided by the system.

2. The DPC interfaces to the DCG and the DVC to set the target performance level to be achieved and requested by the IEM software.

3. The Dynamic Performance Monitor provides the IEM software with some of the system metrics it needs to decide the performance level required.

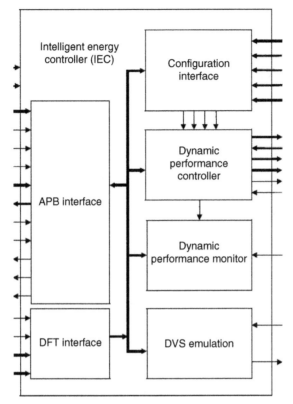

Figure 7.13: IEC Block Diagram [4]

7.6.5 Voltage Islands

In conventional multi-voltage approaches, a design is partitioned into voltage islands that use the lowest supply voltage that allows the maximum performance required by the island. Specifically, one or more voltage islands operate with lower supply voltages and run at correspondingly lower clock frequencies. Each island's voltage and frequency are fixed. In some dynamic voltage/frequency-scaling designs, an island's supply level and clock frequency can change, but only over a few fixed choices that designers set ahead of time.

An adaptive voltage/frequency-scaling approach, such as ARM IEM technology, offers much more flexibility in scaling voltages and frequencies on multiple islands. For this reason, an IEM system is better able to scale performance to meet task requirements. For example, power (and hence energy) is wasted when a processor executes a task quickly but then idles while waiting for the next task. To be truly efficient, the processor should never run faster than is necessary to meet application software deadlines, especially if the time saved on one process is spent idling/waiting for the next process. To help to ensure power and energy efficiency, IEM software policies determine the appropriate task durations for an application and set percentages of full voltage and clock frequency accordingly, to reduce idle states.

IEM technology includes both hardware and software for adaptive adjustment of voltage and frequency to suit processor workload. IEM software collects data directly from the OS kernel, eliminating the need for any special applications encoding to handle this task. IEM software uses various policies to categorize application workload and to profile performance requirements on the fly. IEM software can then set suitable OP for voltage and frequency that enable the processor to meet performance requirements while minimizing idle time. To implement IEM technology, an advanced design methodology and low power physical IP supporting multi-voltage and multi-threshold design are required. The physical IP must be characterized at multiple voltage points, include level shifters and isolation cells, and be available with multiple threshold versions to support leakage power optimization.

7.7 National Semiconductors: PowerWise® Technology

7.7.1 PowerWise Technology

National Semiconductor's PowerWise technology is a system-level approach that reduces the power consumption of SoC solutions used in portable devices. PowerWise technology is a three-part solution: embedded intelligence in digital processors such as the baseband or applications processors of a mobile phone; an open standard power management interface; and companion PMICs.

SoCs use multiple processor and hardware accelerators to provide the processing power required by the applications running on the system. Each of the separate processing engines requires dedicated power management and control to optimize its power consumption. PowerWise technology is suited for managing multiple independent processing engines inside a SoC either when fully operational, or when functions are idling, dormant or completely turned off. The technology creates closed-loop systems where the power-consuming SoC's and power-delivery systems operate in close cooperation, minimizing demands on the power source while providing peak energy efficiency.

7.7.2 Adaptive Power Controller

In addition the technology embeds a synthesizable AMBA-compliant core, the Advanced Power Controller (APC) into the target SoC, as shown in Figure 7.14. The APC ensures

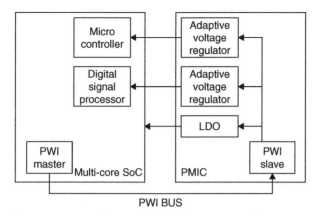

Figure 7.14: Multi-Core SoC and PWI Compliant PMIC

that power-consuming digital logic and power-delivery systems operate in close cooperation, minimizing demands on the power source while providing peak energy efficiency. The APC uses a variety of advanced power management algorithms to achieve this, including adaptive and DVFS. The APC is responsible for monitoring and adjusting the supply voltage of the SoC so that the supply voltage is always optimized for the current operating frequency.

PowerWise technology further addresses the needs of powering microprocessor cores with the ability to implement threshold scaling. With deep sub-micro technology, the static power dissipation due to leakage current becomes significant. Threshold scaling reduces static leakage current by offsetting the N- and P-well bulk biases so that transistors are more effectively driven "off." The APC interfaces to the rest of the system using three interfaces: The AMBA compliant host interface, the CMU interface and the open standard PowerWise Interface (PWI). The host interface is used to control and configure the APC2 while the CMU interface is used to coordinate voltage and frequency changes.

The PWI is used to communicate power management information to external power management ICs to adjust supply voltages. The APC enables the system to implement either dynamic voltage scaling (DVS) or fully adaptive voltage scaling (AVS) on the target system. The use of standard interfaces enables the APC to be easily embedded into any logic circuit and interfaced with other parts of the system. This enables system designers to very quickly develop platforms with next generation power efficiency, minimizing both time-to-market and risk (Figure 7.15).

PWI facilitates adoption of advanced power management techniques that can be used to monitor performance and control various processor voltages such as supply voltages,

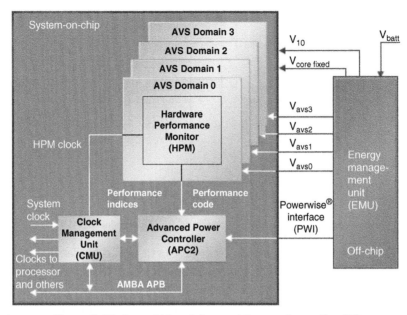

Figure 7.15: PowerWise Advanced Power Controller [5]

threshold voltages, etc. The PWI specification defines a communication interface between a processor and one or more external PMIC. PWI does not include OS interfaces for performance monitoring and control.

7.7.3 The PWI Specification

The demand for advanced power management techniques is driven by the increased functionality in portable devices. To provide longer battery life a number of power management techniques are available:

- Dynamic voltage and frequency scaling

- Adaptive voltage and frequency scaling

- Back-biasing for static leakage reduction

These power management techniques require a link between the SoC and the PMIC that permits fast data transfer and rapid PMIC response. In addition, this interface must be easy-to-use and add little cost.

In 2003 ARM, National Semiconductor and six other companies released the PWI 1.0 standard to support the use of advanced power management techniques in SoCs used in mobile battery operated devices.

The PWI specification defines a two-wire serial interface connecting the integrated power controller of an SoC processor system with a PMIC voltage regulation system that allows system designers to dynamically adjust the supply and threshold voltages on digital processors.

The PWI specification defines the required functionality in the PWI-slave; the operating states, the physical interface, the register set, the command set, and the data communication protocol for messaging between the PWI-master and the PWI-slave. The PWI command set includes PMIC operating state control; register read, register write, and voltage adjust commands.

The PWI standard, PWI 2.0, address new and growing power management issues of complex SoC with multiple processors. The scope of PWI 2.0 covers:

- Advanced power management techniques needed to reduce system power consumption; both dynamic and static leakage power are important

- Multi-processor and multi-core systems

- Advanced hardware and software power management techniques
  - Accurately monitor and control performance level for a given workload
  - Accurately control various supply voltages based on performance level
  - Enable only those functions needed at any given moment
  - Implement system level energy control in real time

- Rapid deployment of advanced power management techniques at the system level requires standardization in hardware and OS interfaces

- PWI 2.0 is a hardware interface standard for advanced system-level power management

PWI 2.0 specification defines a two-wire serial interface dedicated to deployment of advanced power management technologies. In addition it connects one or two SoC processor systems with external PMIC. The PWI 2.0 specification defines the physical interface layer, the register set, the command set, and the data communication protocol for messaging between PWI-masters and PWI-slaves.

Its command set includes PMIC operating state control, register read, register write, and voltage adjust commands as well as master-to-master communication protocol. Also the specification additionally provides provision for user-defined registers in the PWI-slaves and masters.

7.7.4 PowerWise PMU/EMU: Power/Energy Management Unit

A PWI™ 2.0 compliant Energy Management Unit (EMU) for reducing power consumption of low power hand held applications such as dual-core processors and DSPs. A typical EMU (same as a PMIC) contains 2 advanced, digitally controlled switching regulators for supplying variable voltages to a SoC or processor. The device also incorporates five programmable low-dropout, low noise linear regulators for powering I/O, peripheral logic blocks, auxiliary system functions, and maintaining memory retention (dual-domains) in shutdown mode.

The device is controlled via the high speed serial PWI 2.0 open-standard interface. It operates cooperatively with PowerWise® technology compatible processors to optimize supply voltages adaptively (adaptive voltage scaling (AVS)) over process and temperature variations. It also supports dynamic voltage (DVS) using frequency/voltage pairs from pre-characterized look-up tables (Figure 7.16).

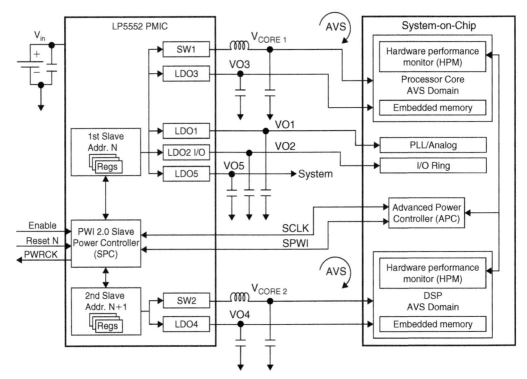

Figure 7.16: LP 5552 EMU/PMU Block Diagram

Source: www.nsc.com

7.7.5 Dynamic Voltage Scaling

DVS exploits the fact that the peak frequency of a processor implemented in CMOS is proportional to the supply voltage, while the amount of dynamic energy required for a given workload is proportional to the square of the processor's supply voltage [6]. Reducing the supply voltage, while slowing the processor's clock frequency, yields a quadratic reduction in energy consumption, at the cost of increased run time.

Completing a task before its deadline is an inefficient use of energy [7]. Employing performance-setting algorithms to reduce the processor's clock frequency only when it is not critical to meet the application's deadlines is critical for energy conservation. There is a significantly lower total energy consumption using DVS compared with traditional gated-clock power management, for the same workload. Note that with DVS, the lower supply voltage reduces static power even when the clock is gated off.

DVS solutions offer improved performance by reducing the supply voltage as the clock frequency is reduced. Open-loop DVS, as shown in Figure 7.17, allows the processor to set the supply voltage based on a table of frequency/voltage pairs. This table must be determined by characterization to assure sufficient margin for all operating conditions and process corners.

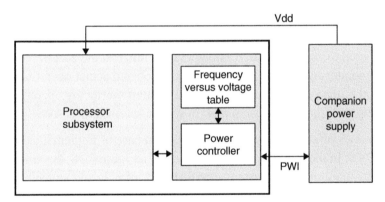

Figure 7.17: Open Loop DVS [8]

The processor must determine the desired operating frequency, request a new voltage, wait for the voltage to stabilize, and then switch itself to the new frequency. The switch may be made immediately when changing from a higher frequency to a lower frequency. When switching from a lower frequency to a higher frequency, the power supply voltage must be high enough to support the new frequency prior to changing the clock.

Power supply stability can be assured either by a time delay or by an analog measurement. Use of a time delay is risky, since there will always be a desire to implement the minimum possible delay for enhanced processor response time.

Open-loop operation is simplified by creating an APC module to off-load the voltage scaling and clock management from the processor. The APC approach supports a common software API allowing the DVS function to be easily accessed by applications or the operating system. Providing a standard interface to the external power supply also simplifies system design and facilitates second-source options for the power supply component. An architecture using an APC is shown in Figure 7.17.

7.7.6 Adaptive Voltage Scaling

AVS is a technique for determining the minimum supply voltage for a digital processor during operation. This technology significantly reduces the supply voltage of a digital processor by removing voltage margins associated with the effects of process and die temperature variation inside the SoC, as well as IR-drop and regulator tolerances in the system.

Performance-setting algorithms optimize power consumption based on workload variations. However, significant power efficiency can also be gained if the SoC does not have to operate under worst-case assumptions but can tune its operating parameters to temporal environmental conditions [9]. SoCs are designed to operate reliably over a wide range of temperature levels and variations of the silicon substrate. Increased voltage levels must be used to assure the large safe operating range at the cost of reduced power efficiency. By monitoring the margin between expected and actual operating conditions, the voltage level of the processor can be reduced without sacrificing operational stability. This closed-loop monitoring of system margin will be referred to as AVS.

Closed-loop or AVS offers improved performance and ease of implementation compared to open-loop DVS. In the closed-loop system shown in Figure 7.18, the voltage is set automatically by monitoring the system's performance margin and adjusting the supply voltage adaptively.

Using a performance request, an APC can set the correct operating voltage and interface with the clock-management unit to enable transitions to new frequencies. The APC receives commands from the Hardware Performance Monitor (HPM) when new, higher frequencies are to be deployed and enables the new clock frequencies on the CPU core and, if applicable, in on-chip cache, memory, and peripherals of the SoC. The HPM can require new voltage levels and fine-tune them by communicating to the external power supply.

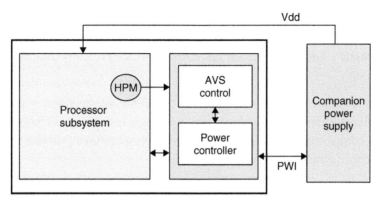

Figure 7.18: Adaptive Voltage Scaling [8]

While the core is still operating under its previous stable frequency, the next higher test frequency is sent to the HPM, its results checked again and again. Voltage is increased in steps until the HPM reports that the test frequency yielded a correct result by issuing a vdd_ok signal to the APC.

The test frequency yields a go/no-go answer which must be augmented by additional HPM logic that can deliver voltage adjustments as the chip's temperature rises and IR drops change, owing to changing demands in supply current. Feedback to the external power supply is based on a local spot on the chip only and may require several sensors for the tightest voltage performance. Downward frequency shifts are less problematic, since the lower frequencies are supported by higher voltages. The combined function of HPM and APC circumvents differences in process, fabrication, and temperature condition.

Since the system is closed-loop in nature, a much finer degree of control over the voltage is possible when compared to the discrete table values in an open-loop system. Response time of the AVS system can be much faster, since it is limited only by the external power supply. The performance measuring circuitry can be used to verify the power supply stability to offer the fastest possible switching from one clock frequency to the next.

Closed-loop AVS is distinguished from open loop, table-based voltage scaling techniques in that it regulates the propagation delay margin in logic cells. In this system, the power supply voltage is a variable that increases or decreases, and the delay margin is a fixed parameter that is regulated over parts, temperature, and clock frequency.

Many advantages arise from this methodology. Closed-loop AVS relaxes the characterization process. There is no need for characterizing voltage/frequency tables because a delay margin is maintained by the AVS feedback loop. Another incentive is that less demand is placed on power supply regulation. The AVS loop adjusts the supply voltage as necessary, compensating for the ±5% tolerance typically allocated to power supply regulation.

By and far the most beneficial advantage is that the minimum operating voltage is realized for all conditions, and can dynamically change as conditions change. Below are the results comparing DVS with AVS at varying temperatures [10] that indicate significant energy savings with AVS (Figure 7.19).

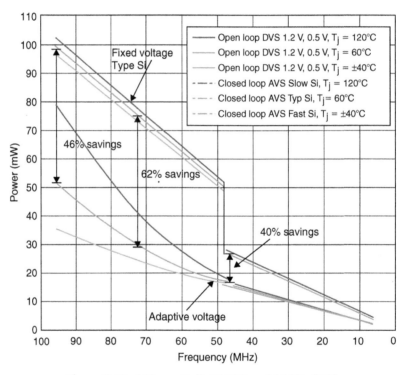

Figure 7.19: 130 nm Bulk CMOS at 96 MHz [10]

7.8 Energy Conservation Partnership

ARM and National Semiconductor have developed energy conservation solutions in order to help mobile device manufacturers to maximize the battery life of their handheld, battery-powered devices. The modularized nature of the total solution means that the technology can be adapted to suit to the underlying performance scaling hardware, including DVS and AVS.

The IEM prediction software determines the lowest performance level that the processor can run at whilst ensuring, with the aid of the IEC, that software deadlines are never missed.

The APC works with the external EMU using the performance prediction to bring the processor to the absolute lowest possible voltage and frequency that still operates the application software correctly (Figure 7.20).

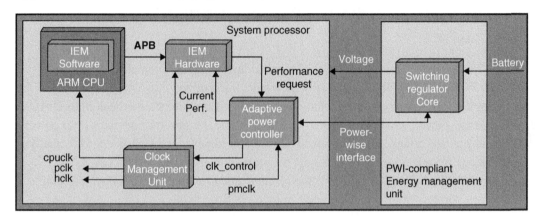

Figure 7.20: ARM and NSC Energy Conservation Solution [11]

This complete solution reduces the energy consumed by the processor to the lowest possible given the constraints of the clock generator and power supply dynamics and the headroom available in the mix of application software.

The ARM IEM technology can be used to reduce the energy requirement of an embedded processor by up to 75%. National's PowerWise technology can reduce safety margins and provide additional energy savings of about 45% using AVS at room temperature as compared with an open loop voltage control solution.

Open-loop and closed-loop voltage scaling are only one area in the project ARM and NSC have undertaken. Clock gating, new cell libraries to minimize leakage current and power domains to turn it off are additional areas where the two companies are collaborating to reduce consumed energy.

7.9 Texas Instruments: SmartReflex

TI's SmartReflex [12] technologies take system-wide perspective on the interrelated issues of power and performance to addresses both dynamic and static leakage power.

SmartReflex technologies are comprised of three facets:

1. Silicon intellectual property (IP).

2. Techniques that can be applied at the SOC design level.

3. System software that manages many of the hardware-enabled SmartReflex technologies, which interface to other power management techniques, based in operating systems (OS) or third-party software subsystems.

7.9.1 Silicon IP

At the silicon level, TI has a track record as an industry pioneer in sophisticated power and performance capabilities, many of which have transitioned into SmartReflex technologies. One major emerging challenge addressed by SmartReflex technologies is static leakage power, which becomes a significantly greater part of a device's total power at smaller process nodes. Several SmartReflex technologies can be applied to drastically limit leakage from a device. For example, the static power leakage in the OMAP2420 processor is reduced by a factor of 40 with SmartReflex technologies. Today, many of TI's 90-nm wireless components already implement SmartReflex technologies to reduce leakage power. In the future, all new devices at the 90 nm, 65 nm and smaller process nodes will incorporate these breakthrough technologies.

Another SmartReflex technology at the silicon level is a library of power management cells that enable power switching, isolation, and voltage shifting to facilitate a granular approach to partitioning a device's power domains. By structuring the device with multiple power domains, functional blocks can be powered down or put into a standby power mode where they are not active, thus reducing power consumption while ensuring optimal performance. To simplify chip-level integration, SmartReflex technologies are supported by an easy-to-use, non-intrusive design flow:

*   *Retention SRAM and logic*: SRAM and logic retention cells support dynamic power switching (DPS) without state loss, lowering voltage, and reducing leakage.

*   *Dual-threshold voltages*: Higher threshold for lower leakage and lower threshold for higher performance.

*   *Power management cell library*: Switching, isolation, and level shifters support multiple domains in SOC implementations.

*   *Process and temperature sensor*: Adapts voltage dynamically in response to silicon processes and temperature variations.

- *Design flow support*: Complete, non-intrusive support for easily integrating SmartReflex technologies.

7.9.2 System-on-Chip

SmartReflex technologies include techniques at the architectural level of SoC design to address static leakage power. These include the following:

- *Adaptive voltage scaling (AVS)*: Maintains high performance while minimizing voltage based on silicon process and temperature.

- *Dynamic power switching (DPS)*: Dynamically switches between power modes based on system activity to reduce leakage power.

- *Dynamic voltage and frequency scaling (DVFS)*: Dynamically adjusts voltage and frequency to adapt to the performance required.

- *Multiple domains (voltage/power/clock)*: Enables distinct physical domains for granular power and performance management by software.

- *Static leakage management (SLM)*: Maintains lowest static power mode compatible with required system responsiveness to reduce leakage power.

7.9.3 System Software

At the level of system software, SmartReflex technologies include host processor power management, which features several capabilities that are deployed at the system level, such as a workload monitor, workload predictor, resource manager, and device driver power management software. Additionally, the SmartReflex framework features DSP/BIOS™ power management software.

SmartReflex technologies support multiple cores, hardware accelerators, functional blocks, peripherals, and other system components. In addition, their system-level technologies are open to OS-based and third-party power management software so that a collaborative and cooperative environment with regards to power and performance can be developed.

- *OS support*: Provides an open environment for blending with operating systems. Supports Symbian, and Linux.

- *Software power management framework*: Intelligent control for power and performance management. Transparent to application programs and legacy code. Monitors system activity, not just processor activity.

- *Workload monitoring and prediction*: Determines system performance needs used to make intelligent power and performance management decisions.

- *Policy and domain managers*: Dynamically controls the system, providing the performance needed at the lowest power.

- DSP/BIOS power management.

7.10 Intel SpeedStep

Intel's SpeedStep® Technology provides the ability to dynamically adjust the voltage and frequency of the processor based on CPU demand [13]. Wireless Intel SpeedStep Technology incorporates three low power states deep idle, standby, and deep sleep. Employing the Wireless Intel SpeedStep Power Manager Software, to intelligently manage the power and performance needs for the end user.

The Power Manager software can be used to manage power consumption and optimize system standby time and talk time in smart phones and other mobile devices. The software provides device programming interfaces (DPIs) to device drivers and applications programming interfaces (APIs) to applications. The platform-specific layer of the Power Manager software is used to adapt the Power Manager software to a given platform.

Device drivers must be set up as a client of the Power Manager software so that the device receives notifications on power policy changes and power states via the DPIs. Optionally, applications can use the APIs to further enhance power savings.

7.10.1 Usage Modes

The Power Manager software has the following usage modes:

- Standby mode (standby time)
- Voice communications (talk time)
- Data communications
- Multimedia (audio, video, and camera)
- Multimedia and data communications (video conferencing)

For each of these usage modes, the Power Manager software provides:

- Optimal power policy for dynamic scaling of power and performance
- Optimal operating frequency and voltage

- Usage of low power modes for the entire system including all of the devices

- State transition and power management for its devices

The Power Manager software changes the operating profile of a generic system as shown in Figure 7.21. DVM and DFM are used to dynamically scale the "Run" frequency and voltage to meet immediate performance requirements with minimum power consumption. New power states are used to minimize "Idle" power consumption.

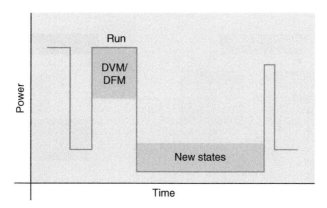

Figure 7.21: Operating Profile with DVM, DFM, and New States [13]

7.10.2 Power Manager Architecture

The power manager architecture consists of the following five software components:

1. The *Policy Manager* is responsible for determining a system's power policy. The Policy Manager uses dynamic scaling of frequencies and voltages to help to provide the lowest power consumption under all types of workloads. The Policy Manager assesses information supplied by the Idle Profiler, Performance Profiler, User Settings, DPIs, and APIs. The Policy Manager defines power states that run the processor at the frequencies and voltages that are consistent with the lowest power consumption.

2. The *Idle Profiler* monitors parameters that are available in the operating system's Idle thread. The Idle Profiler then provides status information to the Policy Manager.

3. The *Performance Profiler* uses the Performance Monitoring Unit (PMU) within the processor to determine if the workload is CPU bound, memory bound, or

both CPU and memory bound. The Performance Profiler then provides status information to the Policy Manager.

4. The *User Settings* allow the user to specify the parameters that are used by the Policy Manager to determine a system's power policy.

5. The *Operating System Mapping* allows the Power Manager software to be ported across multiple operating systems.

The PM software architecture with the five components is shown in Figure 7.22

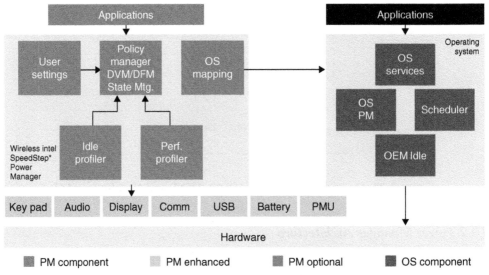

Figure 7.22: Power Manager Software Architecture [e]

7.10.3 Speedstep DFM

The processor's core and peripheral clocks are derived from Phase Locked Loops. The processor implements Intel DFM by allowing the core clock to be configured dynamically by software.

The core clock frequency can be changed in several ways:

- Selecting the 13-MHz clock source

- Changing the core PLL frequency

- Enabling or disabling turbo mode or half turbo mode

7.10.4 Speedstep DVM

The processor implements DVM through its Voltage Manager. The Voltage Manager provides voltage management through use of an I2C unit that is dedicated to communication with an external PMIC regulator, and through use of a Voltage Change Sequencer.

When software initiates a voltage change mode, the Voltage Change Sequencer can automatically send commands via the I2C unit to an external PMIC regulator. The sequencer can send up to 32 commands, which can be categorized as dynamic commands and static commands:

- Dynamic commands are executed when the core is running.

- Static commands are executed after clocks to the processor are disabled.

Intel DVM augments DFM by enabling code to dynamically change the system voltage as well as frequency. Like many other leading companies, Intel's approach to energy conservation is holistic and includes process technology, SoC, and systems development.

7.11 Transmeta LongRun and LongRun2

For its Crusoe chip [14], Transmeta has introduced LongRun, a table-based set of multiple frequency–voltage points, similar to open loop DVS, which helps Crusoe track workloads more efficiently than by using a few steps.

Transmeta's Crusoe processor saves power not only by means of its LongRun Power Manager but also through its overall design which shifts many of the complexities in instruction execution from hardware into software. The Crusoe relies on a code morphing engine, a layer of software that interprets or compiles x86 programs into the processor's native VLIW instructions, saving the generated code in a Translation Cache so that they can be reused. This layer of software is also responsible for monitoring executing programs for hotspots and re-optimizing code on the fly.

As a result, features that are typically implemented in hardware are instead implemented in software. This reduces the total on-chip real estate, capacitance and its accompanying power dissipation. To further reduce the power dissipation, Crusoe's LongRun modulates the clock frequency and voltage according to workload demands. It identifies idle periods of the operating system and scales the clock frequency and voltage accordingly.

Transmeta's later generation, Efficeon processor, contained enhanced versions of the code morphing engine and the LongRun2 Power Manager [15]. LongRun2 includes techniques to reduce static leakage by dynamically adjusting a processor's threshold voltage.

LongRun2 is a broad solution to leakage problems that are emerging in chips made at the 90- and 65-nm nodes. LongRun2 adjusts threshold voltage on a CPU to virtually eliminate leakage current, taking power consumption in idle mode down from 144 to just 2 mW. The technique depends on both hardware circuitry already in Efficeon and the software.

LongRun2 has all the original LongRun techniques with multiple dynamic frequency and voltages. In addition it added dynamic threshold control and other power reducing technologies.

LongRun2 Vt control provides an approach to body bias that can be applied to standard bulk CMOS process technology. It adds capacitance that reduces body contact and power/ground noise. In addition it avoids body bias distribution in metal layers that may increase the SoC die area.

Body bias voltages can be employed to shift the threshold voltage of the SoC transistors toward the worst-case processing corner, resulting in lower leakage (Figure 7.23).

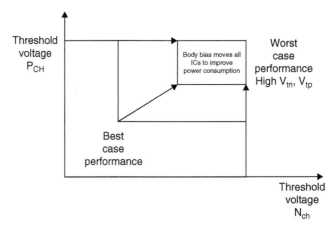

Figure 7.23: Body Bias Employed to Shift Transistor Threshold Voltage of the SoC to Lowest Leakage Corner (Top Right) of the Design Target

A traditional body bias solution can provide some of the advantages of LongRun2 Technologies, however, LongRun2 adds valuable tools and technologies to make body bias more practical to use. In a traditional approach, metal wires are required to distribute the PMOS and NMOS bias voltages across the chip, adding area and routing complexity. LongRun2 adds techniques to simplify the bias distribution reduce or eliminate the need for metal bias routing and allow retrofit of existing designs without a complete re-layout. Besides bias distribution, LongRun2 also offers innovations in dynamic bias control, feedback, and optimization, as well as supporting circuit structures and control techniques that may help to reduce minimum operating voltage.

7.11.1 LongRun2 IP

Transmeta has hardware blocks that semiconductor designers use to quickly implement LongRun2 Technologies into either existing SoC designs or new designs. There are four major categories of supporting blocks for LongRun2 Technologies.

7.11.2 Body Bias Controllers

The design of the threshold voltage controller is critical to achieve the greatest power, performance, and yield benefits. Transmeta is developing a range of controller designs that can be quickly integrated into existing or new designs to provide both yield distribution advantages and low power consumption. These controllers determine the combination of supply and body bias voltages that minimize power dissipation at a target frequency and activity. Advanced controller designs may monitor other factors such as the temperature of the device and degradation due to aging.

7.11.3 Body Bias Voltage Distribution

Traditional body bias voltage distribution requires semiconductor designers to utilize metal layers to route PMOS and NMOS bias voltages. Routing bias complicates global routing by introducing yet another global voltage supply pair, and also results in increased area. With Transmeta's proprietary body bias distribution semiconductor structures, it is possible to retrofit an existing design for body bias quickly with minimal changes to an existing design, reducing design cost and risk.

7.11.4 Body Bias Voltage Generators

NMOS and PMOS bias voltages may be generated off-chip using conventional regulator techniques or on-chip using dedicated bias voltage generators. Due to the low current drive required, on-chip generators consume very modest area and avoid the need to complicate board design and route bias pins externally. Transmeta is developing a range of bias voltage generator designs to enable designers to incorporate these blocks on-chip quickly and easily.

7.11.5 Monitor Circuits

Chip performance and leakage vary with changes in voltage, temperature, load, age, and other operating factors. Transmeta has developed a comprehensive set of monitor circuits that can provide continuous feedback on critical parameters which the Bias Controller can use to improve bias voltage control decisions and increase the benefits of LongRun2 Technologies.

7.12 Mobile Industry Processor Interface: System Power Management

The Mobile Industry Processor Interface (MIPI) Alliance is an open membership organization that includes leading companies in the mobile industry that share the objective of defining and promoting open specifications for interfaces in mobile terminals. MIPI Specifications establish standards for hardware and software interfaces between the processors and peripherals typically found in mobile terminal systems. By defining such standards and encouraging their adoption throughout the industry value chain, the MIPI Alliance intends to reduce fragmentation and improve interoperability among system components, benefiting the entire mobile industry.

The MIPI Alliance is intended to complement existing standards bodies such as the Open Mobile Alliance and 3GPP, with a focus on microprocessors, peripherals, and software interfaces.

7.12.1 System Power Management

MIPI has introduced System Power Management (SPM) Architectural Framework [16]. The document reflects the basis for future standardization activities of the MIPI Alliance System Power Management Working Group. In addition the document proposes a widely applicable abstract model for system-wide power management, defining a set of functional interfaces over which power management control can be performed. Furthermore, a power management request protocol and format are proposed.

MIPI's intention is to allow enough flexibility to enable creativity in power management, whilst ensuring backwards compatibility with hardware and software modules; and the framework should be extensible and scaleable to allow future power control methods. This allows the SPM Architectural Framework to still add significant value in systems where not all components are MIPI SPM aware, and be the foundation for future power management systems.

7.12.2 Power Management System Structure

A representative system model that provides structure, to the SPM challenges, has been developed. The key elements that represent such a system include:

- Devices and subsystem services
- Power management domains
- Policy managers
- Power management infrastructure
- Power management interfaces
- Protocols
- Clock generation and Vdd supply
- Application-supported and application-transparent power management

Representative System Model, used to explain the SPM Architectural Framework in the following sections, is shown in Figure 7.24.

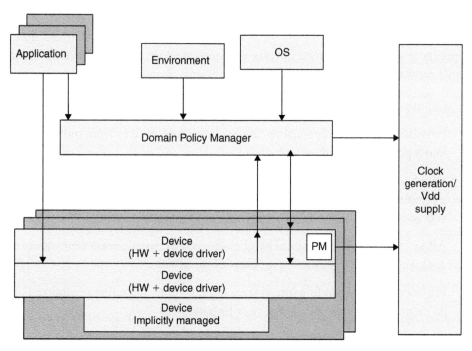

Figure 7.24: Representative System Model of the SPM [16]

The SPM framework identifies categories of device management that have an effect on the system structure and visa versa. These management categories are where policy decisions are made. Categories include:

- Self-managed devices

- Other-managed devices

- Hybrid-managed devices

MIPI SPM architectural framework can be applied to a simple mobile device via a six step process:

1. Power/clock supply partitioning

2. Power management partitioning

3. Policy manager functionality

4. Power management control hierarchy and communication

5. Hardware/software partitioning

6. Device capabilities and CPW/device state mapping

The benefits to system integrators, operating system vendors, software developers, and device IP vendors of the SPM Architectural Framework include:

- Captures power management of whole system in a single framework.

- Provides a structure that enables the integration of devices from multiple sources into a power-managed system.

- Enables device IP vendors to deploy advanced power management capabilities in a portable fashion, and additionally enabling functionality that is decoupled from, and goes beyond that typically provided by different operating SPM models.

- Allows product differentiation by allowing power management innovations to be deployed without redesigning all parts of the system.

- Provides flexibility to fit many system organizations including centralized and de-centralized, hierarchical, and peer-to-peer.

- Provides flexibility to fit many different power management schemes and systems employing a mixture of schemes.

7.13 Summary

Combining high performance with low power consumption is one of the primary goals of portable devices. Historically developers have relied on Doze, Sleep, and Hibernate modes for conserving power in processors. However, an increasing number of system vendors, like Freescale, ARM, Intel, AMD, Transmeta, NSC, Texas Instruments, and others, have developed systems approaches that take advantage of the fact that reducing the frequency and voltage can yield approximately a quadratic decrease in energy used.

A low power system framework, EMS, employs techniques to minimize the drain from the battery or other limited power source. These include:

- The Platform Hardware with various PMICs.

- Performance Estimator employing DVFS, DPTC, and other energy conserving algorithms.

- Performance Setter that maps the estimations to specific platform "operating points" (OP), i.e. combinations of PMC settings.

The EMS dynamically adjusts the performance and power settings of the mobile device to levels that achieve enough performance to handle the task at hand with the minimum power consumed while meeting the real-time deadline.

Freescale has developed an EMS called eXtreme Energy Conservation or XEC. XEC is a framework-based system software targeted at various Freescale chipsets. Software components that comprise XEC include:

- *XEC-DVFS*, which is an OS-aware predictive DVFS that runs on applications processors. Power savings vary widely by use of cases but 40% for video player and audio player software is typical.

- *XEC-LCD*, which automatically regulates the LCD display backlighting uses either or both of two techniques and can save up to 50% of the total LCD power consumption.

The XEC solution has been ported to a number of mobile device platforms to meet the user expectations of high performance wireless and multimedia features with low power consumption.

Using intelligent software algorithms to predict characterize and set the performance and power consumption of wireless and multimedia portable devices, XEC dynamically changes the power states of the system hardware, ensuring the necessary performance required to run applications while minimizing power wastage.

The ARM IEM provides continuous predictive monitoring of the processor demand. Its objective is to run the clock frequency at the lowest available value while still completing the task prior to its deadline. The performance level is set by predictive algorithms that are embedded in the operating system kernel to monitor all the processes.

The ARM IEM and NSC PowerWise technology take a systems view to control additional system components beyond the processor. Additional components such as memory controllers and graphics accelerators can also be monitored in order to control their performance level and power consumption. System-wide energy conservation is therefore possible with the combination of ARM's IEM and NSC's PowerWise technologies.

From a standards perspective, the MIPI SPM Working Group has presented the SPM architectural framework for interfaces that can be used to develop and build flexible and efficiently power-managed portable devices.

The SPM Architectural Framework provides a number of benefits to all parties of the portable device development ecosystem. These include capturing power management in a single framework and enabling device IP vendors to deploy advanced power management capabilities in a portable manner.

References

[1] P. Morris, P. Watson. Automated low-power implementation methodology. *Information Quarterly*, Vol. 4, No. 3, 2005.

[2] C. Watts. Intelligent energy manager. *ARM Forum*, October 2003.

[3] P. Uttley. ARM1176JZF-S intelligent energy management (IEM). *ECoFac*, Nice, France, April 6, 2006.

[4] Intelligent Energy Controller, Technical Overview, revision r0p1, July 2005.

[5] H. M. Lee. National Semiconductors PowerWise brief, June 2007.

[6] T. Mudge. Power: A first class architectural design constraint. *IEEE Computer*, Vol. 34, No. 4, 2001.

[7] K. Govil, E. Chan, and H. Wasserman. Comparing algorithms for dynamic speed-setting of a low-power CPU. *Proceedings of the First International Conference on Mobile Computing and Networking*, San Francisco, CA, November 1995.

[8] R. H. Liu. How to create designs with dynamic/adaptive voltage scaling. *ARM Developers Conference*, Santa Clara, CA, 2004.

[9] S. Dhar, D. Maksimovic, and B. Kranzen. Closed-loop adaptive voltage scaling controller for standard-cell ASICs. *Proceedings of the 2002 International Symposium on Low-Power Electronics and Design (ISLPED 2002)*, Monterey, CA, August 2002.

[10] PowerWise Technology, Powering Next-Generation Portable Devices, National Semiconductors, http://www.nsc.com, May 2004.

[11] C. Watts and R. Ambatipudi. Dynamic management of energy consumption in embedded systems. *Information Quarterly*, Vol. 2, No. 3, 2003.

[12] B. Carlson. SmartReflex™ Power and Performance Management Technologies. White Paper, 2005.

[13] Wireless Intel SpeedStep®Power Manager. White Paper, http://www.intel.com, 2004.

[14] M. Fleischmann. LongRun™ Power Management, January 2001.

[15] http://www.transmeta.com/tech/longrun2.html.

[16] MIPI Alliance SPM Architectural Framework. White Paper version 1.0, http://www.mipi.org/docs/mipi-spm-framework-wp-2005.pdf

CHAPTER 8

Future Trends in Power Management

The power management spectrum covers process, design, architecture, and software and system elements. In the short term the trend in power optimization continues toward a holistic approach. An approach that tightly couples the hardware architecture, operating system, middleware, and application layers to achieve both a high performance user experience and device energy gains.

8.1 Converged Mobile Devices

Many mobile devices today are comprised primarily of discrete electronic systems with mobile computers performing data computations, telecommunications devices providing voice-based communications, and mobile consumer products providing audio, video, and other functions in portable products.

An emerging trend in portable devices is referred to as "converged mobile devices" and is characterized by the convergence of computer, communications, and consumer, product functions into one device. Converged portable devices span consumer, infrastructure, and automotive, aerospace, and biomedical industries. Examples of next generation converged mobile devices include electronic products such as smart phones with cell phone, global positioning system (GPS), sensor, digital wallet, web e-mail access, and medical electronics such as smart medical implants with computing, sensing, imaging, and wireless communication characteristics.

The technologies required to accomplish this convergence of data, audio, speech, video, sensing, and other functions are digital, optical, RF (radio frequency), analog, MEMS (Micro-Electrical and Mechanical Systems), and sensors. Convergence is beginning to take place in the industry typically by discrete and bulky components which do not take advantage of the synergy between the process, System-on-Chip's (SoC) and packaging technology.

Converged applications provide more value, but demand more from batteries. Battery technology is struggling to keep up with processing technology and system requirements.

As an example, carving out the wireless data pipe for many of the future mobile devices. Figure 8.1[1] shows the evolution of cellular mobile systems to 3G or third generation cellular phones.

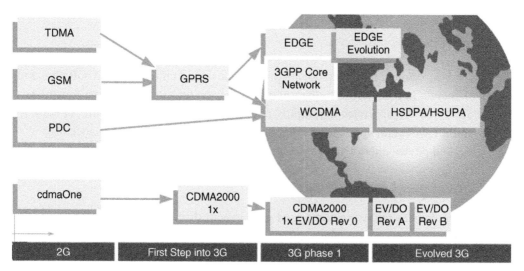

Figure 8.1: Near-Term Evolution of Cellular Mobile Systems [1]

Beyond 3G cellular systems, 4G or fourth generation handsets will support multiple air interfaces for seamless roaming. The goal is to achieve data rates in excess of 100 Mbps. A key technology employed to achieve broadband speeds is orthogonal frequency division multiplexing (OFDM). The change to 4G will improve cellular network efficiency and improved latency and return trip time.

Cellular broadband roadmap, defined by the long-term evolution (LTE) roadmap, with 100 Mbps downlink speed in the 20 MHz spectrum, with OFDM selected and 50 Mbps in the uplink, is illustrated shown in Figure 8.2. LTE offers scaleable bandwidth, reduced latency, and an all Internet Protocol (IP) systems architecture.

Additional broadband wireless technologies will compete with Global System for Mobile Communication (GSM) LTE. These include Ultra-Mobile Broadband (UMB) which evolves from CDMA200 air interface and WiMAX.

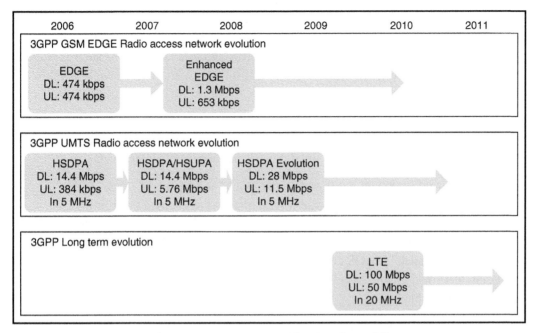

Figure 8.2: Long Range GSM Roadmap [1]

UMB goal is to reach speeds of 280 Mbps where as WiMAX (802.16d) has a theoretical maximum capacity of 70 Mbps.

8.2 Future Processes

Material science will do the heavy lifting required to meet the use-case demands of high performance at low power consumption. From thin body to Fin-FETs, heterogeneous materials including high-k metal gates, MEMS, nanoelectronics, quantum computing, and genetic engineering.

8.2.1 Nanotechnology and Nanoelectronics

Nanotechnology involves the manipulation of materials at the nanometer or the atomic scale to create structures that have novel properties and functions because of their size, shape, or composition. Nanotechnology is a rapidly evolving field of science combining insights from physics, chemistry, biology, material science, and engineering. Nanotechnology combines existing knowledge and ongoing research at the nanometer level, with a specific emphasis on applying the science to engineering materials with enhanced and tailored properties [2].

Nanotechnology is not an industry, but a collection of nanoscale technologies that cut across a broad range of industries and applications. Indeed, nanoscale materials are already commercially used to enhance the properties of basic applications such as water repellant fabrics, antistatic mats, carbon fiber tennis rackets, and sunscreen. In addition, what we call 0.1 μm or 100 nm technology in the manufacture of SoCs is an example of evolutionary nanotechnology that already exists.

The continual downscaling of conventional silicon technology has reached the nanometer scale, and promises to continue this trend down to the near-molecular level.

In addition, this development has provided the "eyes" and "hands" to "see" and to "manipulate" on the nanometer scale which, in turn, has enabled emerging nanoelectronic device technologies. The International Technology Roadmap for Semiconductors (ITRS) shown in Figures 8.3–8.5 has identified trends for both scaled down "classical" CMOS devices and integrated systems, and also for "non-classical" devices and circuit architectures based on more "exotic" nanoelectronic technologies.

Physics and chemistry at the nanometer scale offer new opportunities for electronics and photonics, but it is also well recognized that an integrated approach is needed which includes circuits, devices, and systems.

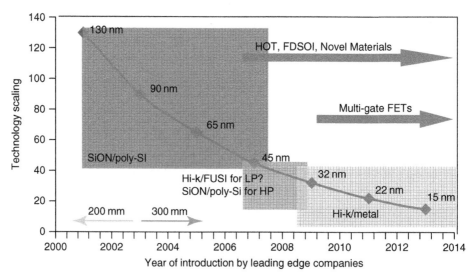

Figure 8.3: Gate Stacks Roadmap [3]

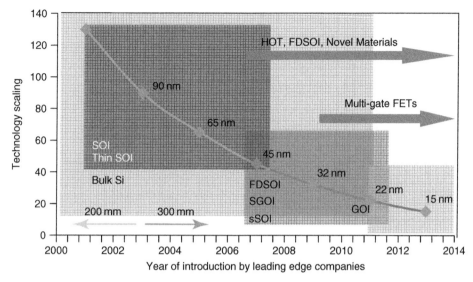

Figure 8.4: Substrate Roadmap [3]

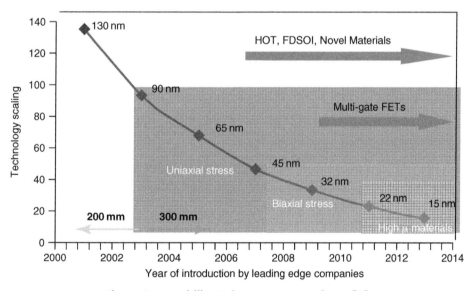

Figure 8.5: Mobility Enhancement Roadmap [3]

Goals for nanoelectronics include:

- Nanometer structures for miniscule transistors and memory ICs that will improve today's computer speed and power consumption by a factor of millions.

- Expansion of mass storage electronics to multi-terabit capacity that will increase the memory storage per unit surface at least a 1,000 times.

- Changes in communication paradigms by increasing bandwidth over 100 times so as to reduce business traveling and commuting.

Regarding nanoscale energy needs, several objectives have been outlined by the National Nanotechnology Initiative energy report including the following:

- Harvesting solar energy with 20% power efficiency and 100× lower cost.

- Solid-state lighting at 50% of the present power consumption.

- Power transmission lines capable of 1 GW transmission.

- Low-cost fuel cells, batteries, thermoelectrics, and ultra-capacitors built from nano-structured materials.

Nanotechnology has already permeated the semiconductor industry, more by gradual evolution than by any revolutionary breakthrough. However, revolutionary changes still lie ahead that include:

- Implementation of single-electron transistors (SETs).

- Nano-imprint lithography.

- Nano-conduction, replacing copper spreader heat sinks for chips with carbon nanotubes.

Figure 8.6 [4] projects out even further in time than the ITRS roadmaps for process technology. Major issues include fewer electrons, high number of interconnects and power consumption.

The need to reduce current leakage, improve battery life, and meet higher performance requirements has led to the adoption of strained silicon, where the lattice constant of crystalline silicon is "stretched" and compressed to achieve higher carrier mobility, and fundamentally increasing the speed at which electrons travel in the transistor. This leads to the implementation of metal/high-k gate structures, which will replace the conventional silicon/oxide transistor gate structure.

While the implementation of metal/high-k gate transistors allows semiconductor manufacturers to deposit a thicker gate dielectric, the deposition of high-k gate materials

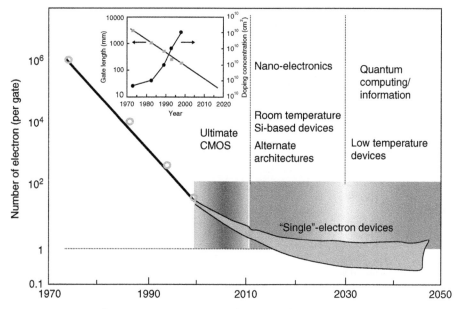

Figure 8.6: Next Generation Material Science [4]

requires a new manufacturing technique, one that deposits a nanometer-thin film one atomic layer at a time. Atomic layer deposition (ALD) deposits controlled thin films one monolayer at a time by alternately pulsing two or more different precursors in the reaction chamber.

ALD technology has matured significantly having cut its teeth in the data storage industry and then migrating to DRAM chip making. With high-k gate dielectric on the horizon for 45 nm, and with potential application at the interconnect level for barrier films, ALD is a by-product of the evolution of nanotechnology in the chip making industry.

Nanotechnology is playing an even larger role than the incremental one evidenced by the advent of strain engineering of the silicon substrate, and metal/high-k gate transistors. Revolutionary changes are ahead, as nanotechnology becomes the next leg to silicon transistor scaling and Moore's law in general, extending this device platform out through 2015 (Figure 8.7).

Emerging devices taxonomy has been defined to give some organization to this growing field of research. The taxonomy consists of four levels: devices, architectures, state variables, and data representations (Figure 8.8).

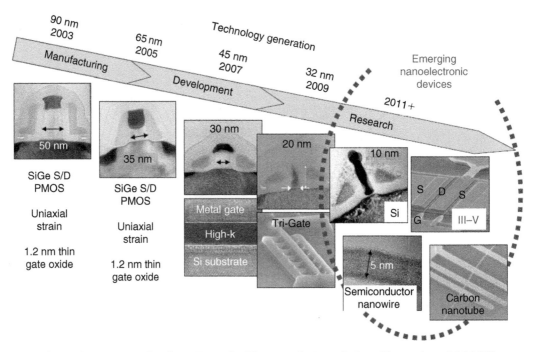

Figure 8.7: Nanotechnology Extends Silicon and Moore's Law Through to 2015 [5]

8.2.1.1 Devices

Devices are the lowest level in the taxonomy. Molecular materials such as carbon nanotubes and silicon nanowires and prototype molecular electronics devices such as field-effect transistors (FETs) are being developed. The material's small molecular structures enable scaling beyond the most advanced lithographic techniques.

Quantum-effect nanoelectronic devices are broken into two groups, namely solid-state nanoelectronic devices and molecular electronics. Single electron transistor (SET) and resonant-tunneling devices are examples of solid-state nanoelectronic devices.

8.2.1.2 Carbon Nanotube

Carbon nanotube devices will drive Moore's law beyond the limits of silicon. Approximately 1 nm in diameter, compared with 35 nm for the physical gate length associated with a 65 nm process technology, a carbon nanotube-based FET has the theoretical potential to run at terahertz clock speeds with a significantly lower power requirement (Figure 8.9).

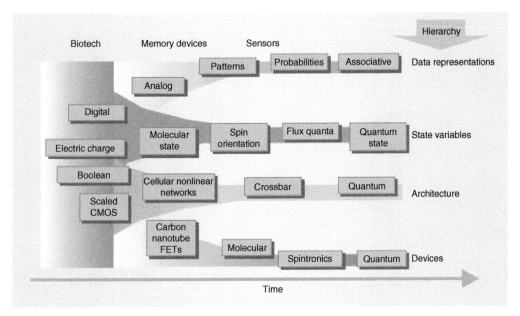

Figure 8.8: Emerging Devices Taxonomy [6]

Carbon nanotubes are formed by taking an atomic planar sheet of carbon atoms bonded together into an array of hexagons and rolled up to form molecular tubes or cylinders with 1 to 20 nm diameters and ranging from 100 nm to several microns in length. Carbon nanotubes have greater ability to carry more current than ordinary metal wires.

Carbon nanowires are the same size as a nanotube. They can be made into transistors, LEDs, and memory structures. Nanotube-based interconnects possess the potential to surmount upcoming resistivity/electromigration challenges with copper, particularly as dimensional shrinks reduce the cross-sectional area required to carry required current and increase the risk of overheating.

Silicon nanowire devices have been fabricated in several geometries. Figure 8.10 shows nanowire devices with silicon nanowire connecting the source and drain contact points.

8.2.1.3 Single-Electron Transistor

A SET is always a three-terminal device, with gate, source, and drain. A SET switches the source-to-drain current on and off in response to small changes in the charge on the gate amounting to a single electron.

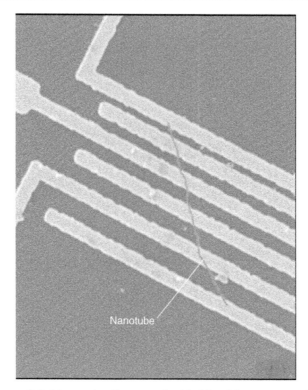

Figure 8.9: Nanotube Compared with Metal Interconnect [7]

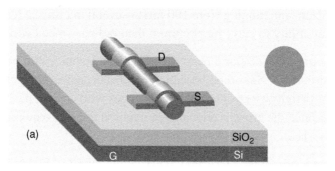

Figure 8.10: Nanowire Device with Silicon Nanowire Connecting Source and Drain [8]

SETs are based around an island, usually of metal, and usually containing a million or more mobile electrons. In order to control the number of electrons on the island, a metal gate electrode is placed nearby. An increase in the voltage of the gate electrode induces an additional electron to tunnel onto the island from the source.

The extra electron soon tunnels off onto the drain. This double-tunneling process repeats millions of times a second, creating a measurable current through the island. Since the current between the source and drain is sensitive to the charge of single electron on the gate, the amplification can be extremely high.

As the gate voltage is increased further, the number of electrons on the island stabilizes at a value one higher than before, and again no current flows. Yet further increases in gate voltage cause more electrons to migrate to the island, and each one-electron increase is heralded by a spike in current flow.

8.2.1.4 Resonant-Tunneling Diode

A resonant-tunneling diode (RTD) is the simplest form of a resonant-tunneling device. A RTD may be a two terminal device without gates. It is made by placing two insulating barriers in a semiconductor, creating between them an island where electrons can reside. A RTDs is made with center islands approximately 10 nm in width. Whenever electrons are confined between two such closely spaced barriers, quantum mechanics restricts their energies to one of a finite number of discrete "quantized" levels. This energy quantization is the basis for the operation of the resonant-tunneling diode.

The only way for electrons to pass through the device is to "tunnel," quantum mechanically, through the two barriers. The probability that the electrons can tunnel is dependent on the energy of the incoming electrons compared to the energy levels on the island of the device.

Adding a small gate electrode over the island of an RTD one constructs a somewhat more complex resonant-tunneling device called a resonant-tunneling transistor. In this three-terminal configuration, a small gate voltage can control a large current across the device. Because a very small voltage to the gate can result in a relatively large current and voltage across the device, amplification is achieved.

8.2.1.5 Molecular Device

Molecular electronics, using individual covalently bonded molecules to act as wires and switching devices, is the longer-term alternative for achieving this increase in density and for continuing Moore's law down to the nanometer scale. Molecular switching devices could be as small as 1.5 nm.

Using either the solid-state or the molecular electronic approach, in order to fulfill the promise of nanoelectronics, it will be necessary to refine greatly both the fabrication techniques and the architectural concepts to permit the useful assembly of small, low power computing structures that contain trillions of nanoelectronic switching devices.

Table 8.1 provides a comparison of the new device types.

8.2.1.6 *Software Is Pivotal*

The implementation of carbon nanotube circuits and nanowires introduces a whole new set of simulation/modeling challenges that require a first-principles approach. Traditional EDA companies will need to combine current software tools with simulation/modeling platforms based on physics at the quantum level.

Modeling and simulation tools [2] for nanotechnology will be required to study sizes scaled from the Quantum to the Mesoscale. Figure 8.11 shows a model example at each length scale.

Moore's law is alive and well. CMOS transistor scaling is expected to continue until around 2020, with new architectures such as tri-gate, carbon nanotubes, nanowires, and III–V materials. Promising research results in areas such as spintronics, phase change, and optical switches will provide a path for beyond 2020.

8.2.2 *Quantum Computing*

A quantum computer is a device which stores information in quantum mechanical two-level systems, called qubits, and exploits fundamental quantum mechanical phenomena to improve computational power. Such devices represent a revolutionary new computational paradigm that has seen tremendous growth following the invention of quantum algorithms with compelling real-world applications, experimental realizations of systems with a few qubits, and the extension of the theory of error correction to quantum systems.

Quantum computers are characterized as something between a device and architecture.

The devices rely on the phase information of the quantum wave function to store and manipulate information. A qubit, the phase information of any quantum state, is extremely sensitive to its external environment. A qubit is easily connected or entangled with the quantum states of particles in the local environment, and no physical system can ever be completely isolated from its environment.

Existing quantum computation experiments with trapped ions, optical interferometers and liquid-state nuclear magnetic resonance (NMR) systems provide a path to understand the fundamentals of practical quantum computing on small scales. However, the future objective of this field is a quantum computer with a scaleable architecture

Table 8.1: Summary of New Devices [8]

Device	Resonant-Tunneling Diode – FET	Single-Electron Transistor	Rapid Single Quantum Flux Logic	Quantum Cellular Automata	Nanotube Devices	Molecular Devices
Types	Three terminal	Three terminal	Josephson junction + inductance loop	Electronic QCA Magnetic QCA	FET	Two terminal and three terminal
Advantages	Density, performance, RF	Density, power, function	High speed, potentially robust (insensitive to timing error)	High functional density, no interconnect in signal path, fast and low power	Density, power	Identity of individual switches (e.g. size, properties) on sub-nm level, potential solution to interconnect problem
Challenges	Matching of device properties across wafer	New device and system, dimensional control (e.g. room temperature operation), noise (offset charge), lack of drive current	Low temperatures, fabrication of complex, dense circuitry	Limited fan out, dimensional control (room temperature operation), architecture, feedback from devices, background charge	New device and system, difficult route for fabricating complex circuitry	Thermal and environmental stability, two-terminal devices, need for new architectures
Maturity	Demonstrated	Demonstrated	Demonstrated	Demonstrated	Demonstrated	Demonstrated

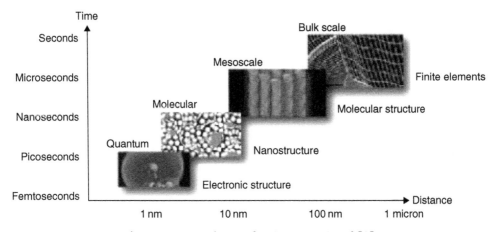

Figure 8.11: Design at the Quantum Level [2]

Source: Accelys

accommodating a large number of qubits, a high clock speed, the capability to perform quantum operations in parallel, and a robust repetitive fabrication method.

Essentially, three different approaches have been taken to the implementation of quantum computers:

1. Bulk resonance quantum implementations

2. Atomic quantum implementations

3. Solid-state quantum implementations

Solid-state systems are the most likely candidates for meeting these demanding requirements. A number of solid-state schemes have been proposed which fall into two categories. The physical qubits are spin states of individual electrons or nuclei, or they are charge or phase states of superconducting structures.

A scheme based on the silicon-based nuclear-spin computer was proposed by Kane who published a breakthrough conceptual study for a silicon-based quantum computer that meets many of these requirements [9]. The key elements of his proposed architecture (see Figure 8.12) are an array of spin-1/2 phosphorus nuclei embedded in silicon and a series of surface gate electrodes.

Kane's solid-state quantum computer has a linear array of phosphorous donor atoms buried into a pure silicon wafer, with an inter-donor spacing of about 20 nm. In the

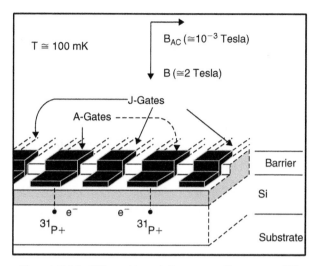

Figure 8.12: Silicon-Based Quantum Computer [9]

presence of a large magnetic field and at sub-Kelvin temperatures, the nuclear spins of the donor atoms can be aligned either parallel or anti-parallel with the field, corresponding to the 0 and 1 qubit states, respectively.

An array of metal gates lies on the surface of the wafer, isolated from the silicon by a barrier layer of SiO2. The A-gates lie directly above the phosphorous donor atoms and enable individual control of single qubits. The J-gates lie between the adjacent donors and regulate an electron-mediated coupling between adjacent nuclear spins, thus allowing for two-qubit operations. Readout of the qubit is performed with either a SET, or with a magnetic resonance force microscope.

Single spin interactions are achieved by changing the voltage on a metallic A-gates electrode positioned above each nucleus. Spin-flips are then carried out by a pulse of RF field tuned to the appropriate Stark-shifted resonance frequency. The electron-mediated interaction between two nuclear spins can be turned on and off by applying a voltage to the electrode placed on the J-gates. Conditional spin-flips can then be achieved again using the RF field. It is possible to show that any desired quantum state can be reached with these two types of interactions.

Building the computer is a daunting enterprise. Individual phosphorus atoms have to be manipulated and placed precisely within a defect-free, isotopically pure silicon matrix. Metal gates created on the nanoscale lay within a few atoms of each other, and each gate is aligned properly over the buried qubits.

8.2.3 Micro-Electrical and Mechanical Systems

MEMS are component-level devices fabricated using semiconductor-like processes (principally etching and plating) that allow for the three-dimensional "sculpting" of silicon and numerous other materials [1]. Most MEMS devices have moving parts, but this is not an absolute requirement. The resulting products are quite diverse, ranging from sensors (inertial, pressure, flow, biological, chemical, and infrared) and actuators (nozzles, pumps, valves, microphones, and switches) to other novel devices, such as lab-on-a-chip and arrays of mirrors.

Most of the MEMS in mobile devices today fall into one of the two categories RF front-end duplexers and microphones. Other types of MEMS devices such as accelerometers for motion detection are beginning to appear in mobile devices. Future applications include navigation aids, input devices, image stabilizers, protection and security, power sources, displays, and biomonitoring.

The adoption of the MEMS solution, rather than some other approach, will be driven by three considerations: cost, physical size, and power consumption.

8.2.3.1 Radio Components

MEMS devices have two key advantages in radio applications for mobile devices: small size and linearity. The small size of MEMS components, and the resulting possibility of packaging them very close to the active circuit elements (perhaps even integrated on the same chip), minimize stray resistance, inductance, and capacitance, resulting in greater range and longer battery life. In addition, since MEMS devices tend to be more linear than semiconductor devices, replacement of a semiconductor component with a MEMS component can reduce signal distortion.

The specific applications for MEMS devices in the RF section of mobile handsets include switching, filtering, matching, and signal generation (oscillators).

Radio sections are the least integrated of the major mobile device sections. This is an advantage for MEMS because it is possible for designers to introduce MEMS devices as one for one replacement for specific parts without a need to completely redo multi-function integrated circuits. It also allows the possibility of changing designs at the board level to gain the full benefits of MEMS.

The main advantages of MEMS implementations of RF functions are reduced signal loss, higher signal isolation, and improved linearity. However these must translate to benefits in size, power consumption, and/or cost if MEMS solutions are to prevail.

8.2.3.2 Power Generation

Fuel cells are sometimes proposed as the solution to the mobile handset power problem and some approaches to fuel cell design involve MEMS. But the fuel cell principles have been known for over 100 years, and, despite their much-hyped advantages, particularly as "clean" power sources for automobiles, fuel cells have not yet entered the mainstream.

At the risk of over simplification, we can define a fuel cell as a device that combines oxygen and hydrogen, in the presence of a catalyst, to create electrical power, heat, and water. The oxygen can come from the air. The hydrogen can come from any one of several sources – a common one, in experimental fuel cells, being a canister of methanol, pressurized for space efficiency.

This gives rise to one of the problems fuel cells face as power sources for mobile handsets. Methanol canisters, even pocket sized ones, are as welcome in the passenger cabin of commercial aircraft as the butane cigarette lighters that they resemble. The hydrogen that the fuel cell extracts from the methanol is itself explosive, as was demonstrated when the Hindenburg Zeppelin blew up causing subsequent airships to adopt helium for lift. A mobile handset that cannot be taken on a commercial flight would not sell in many markets; particularly the high-end markets that can best afford the expected cost premium (at least initially) of fuel cells over traditional battery technology.

It is the need for a catalyst that creates the opportunities for MEMS. Because the power output of a fuel cell is proportional to the surface area of the catalyst, an efficient fuel cell is one that has a high surface area in a small volume. In many designs, the body of the cell resembles a honeycomb with all the surfaces coated with catalyst. MEMS technology can be used to etch high aspect ratio holes in materials, such as silicon, that result in very high surface area to volume ratios.

8.2.3.3 Turbine Engines

Researchers at MIT and the Georgia Institute of Technology are exploring a radical way of using MEMS to solve the energy problem. They are developing prototype millimeter scale versions of the gigantic gas turbine engines that power airplanes and drive electrical generators. These micro-engines will give mobile devices unprecedented amounts of lifetime.

The micro-engines work using similar principles as their large-scale counterparts. They suck air into a compressor and ignite it with fuel. The compressed air then spins a set of turbines that are connected to a generator to generate electrical power. The fuels used could be hydrogen, diesel based, or more energy-dense solid fuels. Made from layers of silicon wafers, these tiny engines are supposed to output the same levels of electrical power per unit of fuel as their large-scale counterparts.

Their proponents claim they have two advantages. First, they can output far more power using less fuel than fuel cells or batteries alone. In fact, the ones under development are expected to output 10–100 W of power almost effortlessly and keep mobile devices powered for days.

As a result of their high energy density, these engines require less space than either fuel cells or batteries. This technology, according to researchers, is likely to be commercialized over the next decade. However, it is premature to know whether it will replace batteries and fuel cells. One problem is that jet engines produce hot exhaust gases that could raise chip temperatures to dangerous levels, possibly requiring new materials for a chip to withstand these temperatures. Other issues include flammability and the power dissipation of the rotating turbines.

8.2.4 Biological (DNA)

Genetic engineering is a powerful and diverse technology that enables biologists to redesign life forms by modifying or extending their DNA. Advances in this domain provide insight into the operating principles that govern living organisms, and can also be applied to a variety of fields including human therapeutics, synthesis of pharmaceutical products, and molecular fabrication of biomaterials, crops, and livestock engineering.

Constructing DNA fragments that consist of almost any gene sequence is not a difficult task. However, the behavior of the resulting genetic constructs is not easy to predict. The first step in making programmed cell behavior a practical and useful engineering discipline is to assemble a component library of genetic circuit building blocks. These building blocks perform computation and communications using DNA-binding proteins, small inducer molecules that interact with these proteins. A component library of cellular gates can be defined that implement several digital logic functions.

Genes can be viewed as nodes in such a network, with input being proteins, and outputs being the level of gene expression. The node itself can also be viewed as a function which can be obtained by combining basic functions upon the inputs (in the Boolean network these are Boolean functions or gates computed using the basic AND, OR, and NOT gates in electronics).

These functions have been interpreted as performing a kind of information processing within a cell which determines cellular behavior. The basic drivers within cells are levels of proteins, which determine both spatial and temporal coordinates of the cell, as a kind of "cellular memory." The gene networks are only beginning to be understood, and it is a next step for biology to attempt to deduce the functions for each gene "node," to assist in modeling behavior of a cell.

Genetic circuit is an approach to model genetic systems using Boolean constructs such as AND, OR, NOT, and NAND. These cellular gates include component for intracellular computation such as NOT, NAND, and devices for external communication. These genetic elements can also be configured to process environmental and internal biochemical analog signals. Table 8.2 describes the mapping of electrical circuits to genetic circuits.

Table 8.2: Electrical Circuits Compared to Genetic Circuits

Electrical Circuits	Genetic Circuits
Basic component of an electronic circuit, transistor	Basic component of a genetic circuit, gene
Binary 1 = high voltage output	Binary 1 = high protein concentration
Binary 0 = low voltage output	Binary 0 = low protein concentration
Outcomes are deterministic	Outcomes are stochastic
Communication occurs in a fixed, closed environment	Communication occurs in an open environment with the signal received by other than the intended recipients

The first step in programming cells and controlling their behavior is to establish a library of well-defined components that serve as the building blocks of more complex systems as shown in Figure 8.13.

8.2.4.1 Biochemical Inverter

In Figure 8.14 [11], the presence or absence of Repressor Protein determines the two possible outputs. In the absence of Repressor Protein, input level = 0, there is a formation of Target mRNA, output level = 1. When the Repressor Protein is present (input level = 1), there is no Target mRNA (output level = 0). The above phenomenon can be modeled by an inverter circuit and defines an example of a Biochemical Inverter.

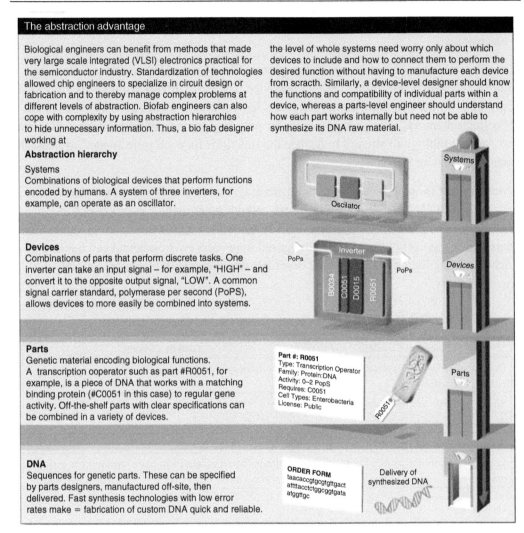

The abstraction advantage

Biological engineers can benefit from methods that made very large scale integrated (VLSI) electronics practical for the semiconductor industry. Standardization of technologies allowed chip engineers to specialize in circuit design or fabrication and to thereby manage complex problems at different levels of abstraction. Biofab engineers can also cope with complexity by using abstraction hierarchies to hide unnecessary information. Thus, a bio fab designer working at the level of whole systems need worry only about which devices to include and how to connect them to perform the desired function without having to manufacture each device from scratch. Similarly, a device-level designer should know the functions and compatibility of individual parts within a device, whereas a parts-level engineer should understand how each part works internally but need not be able to synthesize its DNA raw material.

Abstraction hierarchy

Systems
Combinations of biological devices that perform functions encoded by humans. A system of three inverters, for example, can operate as an oscillator.

Devices
Combinations of parts that perform discrete tasks. One inverter can take an input signal – for example, "HIGH" – and convert it to the opposite output signal, "LOW". A common signal carrier standard, polymerase per second (PoPS), allows devices to more easily be combined into systems.

Parts
Genetic material encoding biological functions. A transcription operator such as part #R0051, for example, is a piece of DNA that works with a matching binding protein (#C0051 in this case) to regular gene activity. Off-the-shelf parts with clear specifications can be combined in a variety of devices.

DNA
Sequences for genetic parts. These can be specified by parts designers, manufactured off-site, then delivered. Fast synthesis technologies with low error rates make = fabrication of custom DNA quick and reliable.

Figure 8.13: Components to Systems [10]

8.3 Future Packaging for Mobile Devices

8.3.1 System Packaging Evolution

Historically packaging plays two roles. First, it provides I/O connections to and from discrete integrated circuits; second, it interconnects both active and passive components on system-level boards. Integrated circuits have begun to integrate not only more

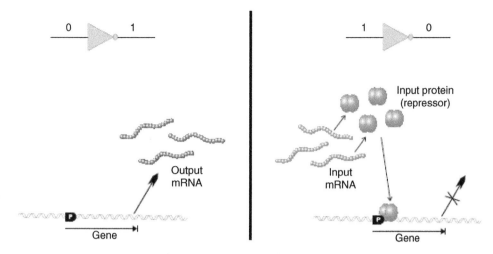

Figure 8.14: Biochemical Inverter

transistors, but also active and passive components on an individual chip, leading to as SoC. However, the disparate process technologies that are required to manufacture digital and analog integrated circuits limit the ability to develop, manufacture, and test highly integrated systems in an economic fashion.

Therefore, the SoCs approach presents fundamental, engineering, and investment limits, as well as computing and communication limits for wireless and wired systems over the long run. In addition the packaging that is used to provide I/O connections from the chip to the rest of the system is typically bulky and costly, limiting both the performance and the reliability of the IC it packages. Systems packaging, involving the interconnection of components on a system-level board, is similarly bulky and costly with poor electrical and mechanical performance.

This led to packaging approaches like multi-chip modules (MCM), system-in-package (SiP), package-on-package (PoP), and future 3D packaging technologies like redistributed chip packaging (RCP) and systems-on-package (SoP). Figure 8.15 shows the evolution of systems packaging.

8.3.2 Redistributed Chip Packaging

RCP eliminates the substrate and the need for intermediate interconnects typically found with IC packaging solutions such as flip chip ball grid array (BGA) and wirebond ball grid array. These technologies depend on the bumps or wires combined with a substrate to provide a complete interconnect solution.

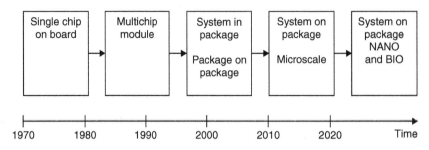

Figure 8.15: Systems Packaging Evolution

By using a dielectric material buildup process [12] (Figure 8.16), the RCP technology creates the required electrical signal interface, provides the signal interconnect solution, while at the same time defining the final package body size.

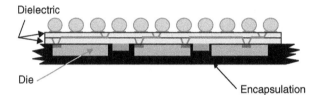

Figure 8.16: Build Up and Metallization Constructed on Embedded Die Employing Plating and Photolithography

In addition to reducing the interconnect complexity and size of the package, RCP technology also improves the overall package electrical characteristics. The low-k dielectric materials are used as interlayer dielectrics to effectively reduce propagation delay, cross-talk noise, and power dissipation due to resistance and capacitance coupling (Figure 8.17).

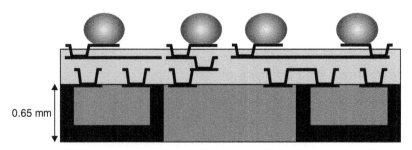

Figure 8.17: RCP Cross-Section [12]

During the buildup process of the RCP technology there is the ability to integrate additional functionality into the package. Power busses required for multi-core designs can be moved from the die design and into the package fabricating process. In a similar fashion, inductors can be built. Circuits such as harmonic filters, multi-layer inductors, and VCOs can be integrated into the package during the fabrication process. In addition, passive components such as chip capacitors and resistors can be embedded at multiple levels within the package. This reduces the number of external components required for a system-level design. This capability is critical for enabling the integration of functionality into small, compact converged wireless and portable devices. The general process flow for RCP is illustrated in Figure 8.18.

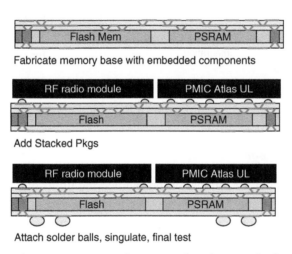

Figure 8.18: General Process Flow for RCP [12]

As illustrated in Figure 8.19, the RCP system integration process for a SiP, takes memory, RF and PA (power amplifier) modules, baseband or applications processors and power management components and integrates them into a single chip.

Figure 8.19 shows an RCP system integration of a cellular radio. This fully integrated system solution is physically the size of a US postage stamp (18 × 20 mm) and includes a 32 MB NOR FLASH and 8 MB PSRAM embedded in the "memory base." Other components include two crystals, a quad-band saw filter, and 26 chip capacitors.

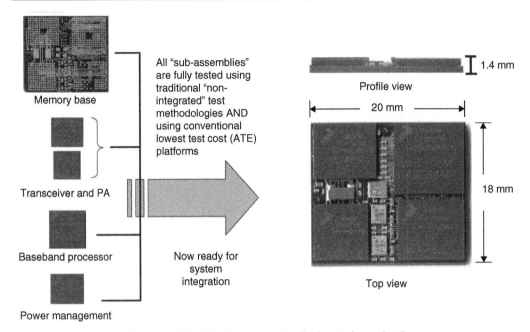

Memory base

All "sub-assemblies" are fully tested using traditional "non-integrated" test methodologies AND using conventional lowest test cost (ATE) platforms

Transceiver and PA

Now ready for system integration

Baseband processor

Power management

Profile view — 1.4 mm

20 mm

18 mm

Top view

Figure 8.19: RCP Enables a Radio-in-Package [12]

8.3.3 System-on-Package

The SoP concept overcomes a number of the engineering limits of a SoC and other systems packaging techniques. As IC integration moves to nanoscale and wiring resistance increases, the global wiring delay in a SOC becomes too high for computing applications. This leads to what is referred to as "latency" which can be avoided by either moving global wiring from the nanoscale on ICs to the microscale on a SoP, or making the digital chips much smaller. A SoP resolves both of these challenges.

Wireless integration limits of a SoC are also handled well by a SoP. RF components such as capacitors, filters, antennas, switches, and inductors are fabricated in the package rather than on silicon.

The SoP concept integrates multiple system functions into one small, lightweight, thin-profile, low-cost, high performance packaged system. The system design may call for high performance digital logic, memory, and graphics, and analog signals for RF and video, as well as broadband optical functions. Unlike a SOC, however, no performance compromises have to be made in order to integrate these disparate technologies since each technology is separately integrated into the SoP package (Figure 8.20).

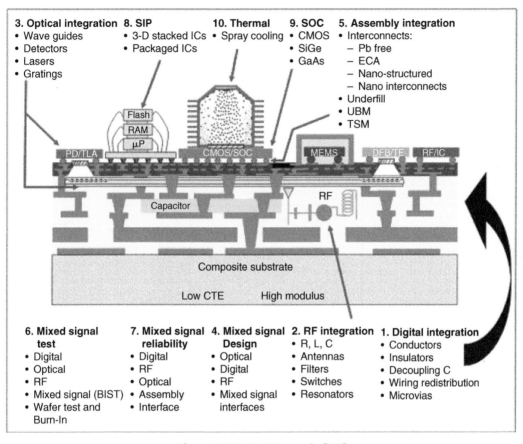

Figure 8.20: SoP Example [13]

RCP and SoP support Moore's law. In the case of SoP, combining nanoscale ICs with newly developed micro- to nanoscale, thin-film versions of discrete and other integrated components. It embeds both of these components in a new type of package so small that it eventually will transform handhelds into multi- or mega-function systems. However, SoP products are being developed not just for convergent electronics that combine computing, communication, and consumer electronics.

8.4 Future Sources of Energy for Mobile Devices

Renewable energy sourcing has grown from long-established concepts, like water wheels and windmills, into devices for powering mobile electronics. Systems can forage for power from human activity or derive limited energy from ambient heat, light, radio, or vibrations. Some examples are shown in Table 8.3.

However, the mobile device industry is investing heavily in Fuel Cell technology in response to the ever-growing appetite of mobile devices and their users.

Table 8.3: Energy Harvesting Opportunities [14]

Energy Source	Performance	Notes
Ambient radio frequency	$<1\ \mu W/cm^2$	Unless near a transmitter.
Ambient light	$100\ mW/cm^2$ (directed toward bright sun) $100\ \mu W/cm^2$ (illuminated office)	Common polycrystalline solar cells are 16–17% efficient, while standard monocrystalline cells approach 20%. Although the numbers at left could vary widely with a given environment's light level, they are typical for the garden-variety solar cell Radio Shack sells.
Thermoelectric	$60\ \mu W/cm^2$	Quoted for a Thermo Life generator at $\Delta T = 5°C^B$; typical thermoelectric generators $\leq 1\%$ efficient for $\Delta T < 40°C$.
Vibrational micro-generators	$4\ \mu W/cm^3$ (human motion – Hz) $800\ \mu W/cm^3$ (machines – kHz)	Predictions for $1\ cm^3$ generators. Highly dependent on excitation (power tends to be proportional to ω^3 and y_0^2, where ω is the driving frequency and y_0 is the input displacement), and larger structures can achieve higher power densities. The shake-driven flashlight, for example, delivers $2\ mW/cm^3$ at 3 Hz.
Ambient airflow	$1\ mW/cm^2$	Demonstrated in micro-electromechanical turbine at 30 liters/min.[2]

(Continued)

Table 8.3: (Continued)

Energy Source	Performance[a]	Notes
Push buttons	50 µJ/N	Quoted at 3 V DC for the MIT Media Lab Device.[2]
Hand generators	30 W/kg	Quoted for Nissho Engineering's Tug Power (versus 1.3 W/kg for a shake-driven flashlight).
Heel strike	7 W potentially available (1 cm deflection at 70 kg/1 Hz walk)	Demonstrated systems: 800 mW with dielectric elastomer heel,[2] 250–700 mW with hydraulic piezoelectric actuator shoes,[2] 10 mW with piezoelectric insole.[2]

8.4.1 Fuel Cells

Wouldn't it be great to never have to pack your mobile device charger again? In fact, what if your mobile device didn't even use a battery, and instead would run several days on a few drops of alcohol?

Fuel cells seem like an excellent power source, cheap to build, requiring no toxic metals, and having the ability to "charge" in a matter of seconds. Even perhaps more amazing is their efficiency given there size and weight. Most micro-fuel cells in development are from two to five times more efficient at producing power for their size than even the super-efficient lithium ion battery.

Although fuel cells have great promise, the idea is nothing new. Sir William Grove first conceived the operating principle of a fuel cell in 1839 in England. In the 1960s, NASA used alkaline fuel cells (AFC) to power space missions. Fuel cell power generating plants with outputs of more than 200 kW are operating today.

Like batteries, fuel cells are devices that convert chemical energy directly into electrical energy. Batteries are designed such that a fixed amount of active chemistry is included in the system and when used up, must be discarded (primary batteries) or recharged (secondary batteries).

Fuel cells differ from batteries in that the active chemistry, or fuel, is held in a separate containment tank and supplied to the electrochemical "engine" when electricity is desired. As long as the supply of fuel is maintained, the device will continue to generate electricity. Theoretically, a very large tank could provide extremely long service; however, as a more practical approach, the fuel is held in smaller, more convenient, replaceable cartridges that can be continuously changed to provide extended run time.

A fuel cell has both an anode and cathode, but unlike a battery these electrodes never change during the chemical reactions that produce power. Instead, a fuel serves as the electron source and the electrodes act as catalysts.

The fuel cell relies on an oxidation/reduction reaction, as with a battery, but the reaction takes place on the fuel rather than the electrodes. As long as the cell receives a supply of fuel, the fuel cell produces electricity. In a fuel cell, the anode is bathed in the fuel, and the cathode collects and makes available the oxidant. An ion-conducting membrane separates the two, allowing the reaction to take place without affecting the electrodes.

So if fuel cells are so clean and efficient, why do not the latest Nokia or Motorola cell phone models sport them, at least as an option? After all, Motorola even sells an accessory crank charger for its phones, why not a fuel cell?

8.4.1.1 Fuel Cell Technology

Fuel cells generate electricity from an electrochemical reaction between oxygen and a fuel like hydrogen, methanol, gas, etc., which combine to form water. There are a variety of different types of fuel cells, though they are all essentially based on a central design, and the electricity produced can be used to power an array of devices, including cars and buses as well as laptops, mobile phones, and an assortment of mobile communications. Heat, a by-product of fuel cells, can also be applied to certain applications, such as keeping a house warm.

What is described as the fuel cell itself consists of a so-called fuel cell stack, which is built up from a number of individual cells. An illustration of the fuel cell stack is illustrated in Figure 8.21.

Each individual cell within this stack has two electrodes, one positive and one negative, called the cathode and the anode. The reactions that produce electricity take place at the electrodes. Every fuel cell also has an electrolyte, which carries electrically charged particles from one electrode to the other, and a catalyst, which accelerates the reactions at the electrodes. The electrolyte plays a crucial role as it must only permit the appropriate ions to pass between the anode and cathode. If free electrons or other substances were able to travel through the electrolyte, the chemical reaction would be disrupted.

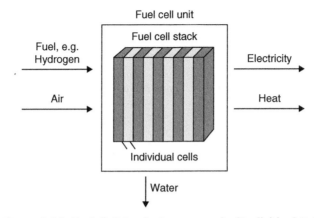

Figure 8.21: Fuel Cell Stack Constructed of Individual Cells

Source: http://www.fuelcelltoday.com

An illustration of this process, whereby the electrolyte plays a key role in permitting appropriate ions to pass between the anode and cathode, is provided in Figure 8.22.

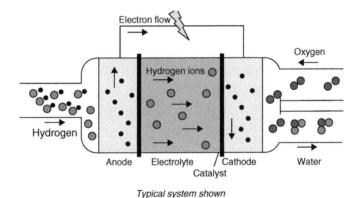

Typical system shown

Figure 8.22: Fuel Cell System

Source: http://www.fuelcelltoday.com

8.4.1.2 Types of Fuel Cells

Fuel cells are classified primarily by the type of electrolyte employed, which determines the sort of chemical reactions that take place within the cell, the type of catalysts required, the temperature range in which the cell operates, and the fuel required, as well as other factors. These characteristics affect the applications for which these fuel cells are most suitable. Several forms of fuel cells are currently under development, each with its own advantages, limitations, and potential applications. Some of the most promising types of fuel cells are summarized in Table 8.4:

Table 8.4: Types of Fuel Cells

• Alkaline
• Direct methanol
• Metal air
• Molten carbonate
• Phosphoric acid
• Protein exchange membrane (PEM)
• Regenerative (reversible)
• Solid oxide
Source: www.fuelcelltoday.com

- *Alkaline Fuel Cells (AFC)*: It uses an alkaline electrolyte, such as potassium hydroxide, and was originally used by NASA on space missions.

- *Direct Methanol Fuel Cells (DMFC)*: A relatively new type of fuel cell, the DMFC is similar to the proton exchange membrane (PEM) cell in that it uses a polymer membrane as an electrolyte. However, a catalyst on the DMFC anode draws hydrogen from liquid methanol, eliminating the need for a fuel reformer. Therefore, pure methanol can be used as fuel.

- *Metal Air Fuel Cells (MAFC)*: These are not fuel cells in a conventional way, but work similarly to batteries by generating electricity using metal and oxygen. However, they can be rechargeable.

- *Molten Carbonate Fuel Cells (MCFC)*: It uses a molten carbonate salt as the electrolyte. MCFC has the potential to be fueled with coal-derived fuel gases, methane, or natural gas.

- *Phosphoric Acid Fuel Cells (PAFC)*: It consists of an anode and a cathode made of a finely dispersed platinum catalyst on a carbon and silicon carbide structure that holds the phosphoric acid electrolyte. This is the most commercially developed type of fuel cell and can also be used in large vehicles such as buses. Many fuel cell units sold before 2001 used PAFC technology.

- *Proton Exchange Membrane Fuel Cells (PEM)*: The hydrogen fueled PEM fuel cell uses a polymeric membrane as the electrolyte, with platinum electrodes. These cells operate at relatively low temperatures and can vary their output to meet shifting power demands. These are the best candidates for cars, buildings, and smaller applications. It is also called a polymer electrolyte fuel cell (PEFC).

- *Regenerative Fuel Cells (RFC)*: This class of fuel cells produces electricity from hydrogen and oxygen, but can be reversed and powered with electricity to produce hydrogen and oxygen; effectively storing energy or electricity.

- *Solid Oxide Fuel Cells (SOFC)*: The cells work at even higher temperatures than molten carbonate cells. These cells can reach efficiencies of around 60% and are expected to be used for generating electricity and heat in industry and potentially for providing auxiliary power in vehicles.

8.4.1.3 Applications

Because most current fuel cells being developed in the lab are rather large by cellular phone standards, it is likely that their first applications for mobile devices will be as chargers for cellular phones with lithium ion batteries rather than the batteries themselves. As fuel cells shrink in size, they will start appearing as a power source rather than a charger. Table 8.5 provides a brief description of each fuel cell technology and potential applications.

When fuel cells arrive for cellular handsets, will most users want one? Most likely no, at least not at first. While fuels do offer the potential of longer run times in handsets and portable devices, so do bigger battery packs. In addition, over time, battery capacity for a given size has slowly been improving, and power consumption of wireless devices has slowly been decreasing.

While it is true that most 3G cellular devices are power hogs, this area has been improving as well. If fuel cells in portable wireless device are to become prolific, they need to be as easy to use as current battery systems. This will be tough, since lithium ion batteries are light, easy to charge, and reliable. Assuming that fuel cells can be made to operate over a wide temperature range, in any position, there still is the fact that they need to constantly be recharged with fuel, and this process must be made extremely easy and inexpensive. For comparison, to charge a typical lithium ion battery at prevailing electric rates costs less than 1/10 of a cent of electricity (800ma/H 3.6v lithium ion battery at an electric rate of $0.10/KW/H).

Table 8.5: Brief Description of Each Fuel Cell Technology and Potential Applications

	AFC	DMFC	MCFC	PAFC	PEMFC	SOFC
Electrolyte	Potassium hydroxide	Polymer membrane	Immobolized liquid molten carbonate	Immobolized liquid phosphoric acid	Ion exchange membrane	Ceramic
Operating temperature	60-90°C	60-130°C	650°C	200°C	80°C	1,000°C
Efficiency	45-60%	40%	45-60%	35-40%	45-60%	50-65%
Typical electrical power	Up to 20 kW	<10 kW	>1 MW	>50 kW	Up to 250 kW	>20 kW
Possible applications	Submarines, spacecraft	Portable applications	Power stations	Power stations	Vehicles, small stationary	Power stations

Source: http://www.fuelcelltoday.com

Fuel cells are a great application for those without a good source of electrical power, or for those that travel a great deal and do not want to carry around a charger. Fuel cell chargers might even appeal to a wider audience, if they could be made inexpensive and convenient. A charger does not have the size and operating position restrictions that a fuel cell battery pack would, and may appeal to some in the high-tech crowd.

8.4.1.4 The Advantages and Disadvantages of Fuel Cells

The Disadvantages
As it turns out, there have been quite a few potholes on the road to super clean handset energy. These problems have not been easy to overcome, and because of this, it will be several more years before we will be seeing an actual product in a cell phone store nearest you.

For one, most micro-fuel cell designs use methanol as their fuel source. While methanol is environmentally friendly and not explosive, the fuel is toxic and flammable. For example, "flammable fluids" are listed as prohibited substances on most commercial airliners. Although not running a cellular phone on an airplane is probably not a problem, not being able to carry one on an airplane definitely would not be acceptable to most.

Still, methanol is not the only fuel that fuel cells can operate on. Others use ethanol but typically they have lower efficiencies. Hydrogen can work as well, but there are many obvious problems associated with that gas. Regular gasoline, as well as a few other common fuels, generally is not a good source of fuel for a fuel cell due to both safety consideration and operational ones.

A second fuel cell problem is waste. Fuel cells generate waste of two kinds, thermal and chemical. The oxidation reaction inside the fuel cell is the same reaction that occurs in a fire, and as with a fire, it is exothermic – that is, it produces heat. Fuel cells run warm.

Those aimed at battery replacements may have internal temperatures in the range of 50–100°C, up to the temperature of boiling water. Some newer fuel cells, however, run at much lower temperatures, even approaching room temperature.

A third potential problem with using fuel cells goes hand-in-hand with the second, that is, they generally prefer to operate at high temperatures, sometimes very high temperatures. This is not good for someone who wants to use a fuel cell for a cellular handset and keep it in his or her pocket. New designs are improving in this area, with some functioning well at room temperatures, but all fuel cells still have a difficult time at really low temperatures in the under 0°C temperature range. Some fuel cells have trouble operating in very dry conditions as well.

If these problems were not enough, there are others. For example, some fuel cells can only operate in one orientation or another. Move these fuel cells the wrong way and they stop operating, not a good characteristic for a portable device like a handset.

Still yet another problem is the fuel cell's operating life. Even when the fuel is replaced, the fuel cell itself wears out over time, just as used batteries do. For this reason, not only does a fuel cell need a means to be filled with fuel, but it also needs a means of being replaced. Typical life for fuel cells is about a year, but the technology is advancing rapidly, and life of a fuel cell should approach that of lithium ion batteries before long.

One final obstacle to be overcome is a fuel cell's output voltage, typically around 0.5 V. A DC-to-DC converter can be used to boost this voltage up to usable levels, but this can add inefficiencies into the system, and can add to the total cost. A solution would be to put several fuel cells in series, but this has problems as well. A fuel cell needs to "breath" and stacking them in any way would restrict their ability to intake oxygen and exhale carbon dioxide and water.

The Advantages

While fuel cells have some definite problems to be worked out, they have many advantages over conventional batteries as well. First, is the obvious one, you are no longer required to charge your handset for several hour everyday. Instead, the fuel cell needs to be filled with fuel by some means. Once this has occurred, the fuel cell will power the device until the fuel is exhausted.

This brings up a second advantage of a fuel cell, the ability to see a fuel level, something not possible with conventional batteries. A fuel cell can have a fuel window much like a pocket lighter. When the fuel is empty, the device will stop functioning. Not all devices will likely have a fuel-level window, just as some butane lighters do not, but the ability to have a fuel window could be a big plus for some applications.

Another advantage of a fuel cell is its ability to supply a relatively large amount of power for a given size and weight device. Certainly, this will be important for future 3G devices which will likely demand even more from a power source than current devices. WCDMA handsets available in Japan give a glimpse of this. Some handset models currently available in Japan have talk-times under an hour, and standby times under a day. A fuel cell could extend this indefinitely with multiple fillings.

The advantages and disadvantages are summarized in Table 8.6.

8.4.1.5 Life with a fuel cell

When fuel cells are integrated into cellular handsets and other mobile devices, what will life with such a device be like? How will it differ from today's devices with lithium ion batteries? Do you want to be the first one on your block with such a device?

When you remove your new fuel cell handset from its box, you will probably notice that your new phone is slightly larger than your current handset. Most micro-fuel cells approaching commercialization are somewhat larger than their lithium ion counterparts, and this will require slightly larger handsets, although the weight of a fuel cell handset likely would not be any higher than that of currently available models.

The next thing that you will notice is that they have not included a plug-in charger in the box, and rather have given you a small fuel pack about the size of a matchbook. This fuel pack, you will learn, is good for 3 or 4 charges of the fuel cell. The instructions included will tell you that additional fuel packs can be purchased for under a dollar each.

Your next task is to "charge" your new phone. You follow the instructions connecting the fuel pack to the handset, and squeezing it until the handset's fuel window shows that it is full. The phone instantly springs to life. Depending on the model of handset or mobile device, it may include a compartment to store an extra fuel pack (Figure 8.23).

Table 8.6: Advantages and Disadvantages of Fuel Cells

Advantages	Disadvantages
• Fuel cells eliminate pollution caused by burning fossil fuels; the only by-product is water.	• Fueling fuel cells is still a major problem since the production, transportation, distribution, and storage of hydrogen is difficult.
• If the hydrogen used comes from the electrolysis of water, then using fuel cells eliminates greenhouse gases.	• Reforming hydrocarbons via reformer to produce hydrogen is technically challenging and not clearly environmentally friendly.
• Fuel cells do not need conventional fuels such as oil or gas. This eliminates economic dependence on politically unstable countries.	• The refueling and the starting time of fuel cell vehicles longer and the driving range is shorter than in a conventional car.
• Since hydrogen can be produced anywhere where there is water and electricity, production of potential fuel can be distributed.	• Fuel cells are in general slightly bigger than comparable batteries or engines. However, the size of the units is decreasing.
• Installation of smaller, stationary fuel cells may lead to a more stabilized and decentralized power grid.	• Fuel cells are currently very costly to produce since most units are handmade.
• Fuel cells have a higher efficiency than diesel or gas engines.	• Some fuel cells use very costly materials.
• Most fuel cells operate silently compared to internal combustion engines.	• The technology is not yet fully developed and few products are available.
• Low temperature fuel cells (PEM, DMFC) have low heat transmission, which makes them ideal for military applications.	
• Operating times are much longer than with batteries, since doubling the operating time requires only doubling the amount of fuel and not the doubling of the capacity of the unit itself.	
• Fuel cells have no "memory effect" when they are getting refueled (i.e. they do not harm a memory).	
• The maintenance of fuel cells is simple since there are few moving parts in the system.	
Source: www.fuelcelltoday.com	

Over the course of the next few days, you will discover that your fuel cell handset seems to perform like any other handset, but its "battery" life is about two or three times longer than you remember your lithium ion phone lasting. After a long call, you notice the fuel cell compartment slightly warm to the touch, but it is not at all uncomfortable.

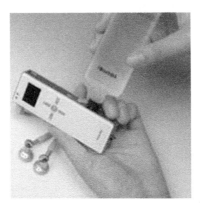

Figure 8.23: Toshiba DMFC-Based MP3 Player

Source: http://www.techsmec.com

When your handset's fuel cell runs low on fuel, it is indicated on the handset's display. You take out your fuel pack, connect it to your handset, and again squeeze until your handset is full. You do not have to wait until your handset is low on fuel to fill it. You can fill it anytime.

Overall, the future of fuel cell power for portable devices looks bright, although it will be a few more years before commercial products start hitting the market for portable wireless devices. The technology is difficult, and a few technical hurdles still need to be worked out before the technology is used on a widespread basis. Currently, in terms of specific technologies the Proton Exchange Membrane Fuel Cell is predicted to be the fastest growing technology, followed by Solid Oxide and Direct Methanol.

8.5 Future Displays for Mobile Devices

Future displays [15] need to distinguish themselves in the area of power consumption, thickness, supporting electronics, size of the image, and viewing image.

8.5.1 Electronic Paper Displays

Electronic Paper Displays, also called ePaper or electronic ink, is one of the most promising technologies for future display. The image looks like ink and paper and has very low power consumption due to bistability and "power-off" image retention. Electrophoretic displays (EPDs) are a technology enabled by electronic ink that carries a charge enabling it to be updated through electronics. Electronic ink is ideally suited for

EPDs as it is a reflective technology which requires no front or backlight, is viewable under a wide range of lighting conditions, including direct sunlight, and requires no power to maintain an image. Some direct view ePaper contenders include iMOD and Electrophoretic.

8.5.1.1 Technology: Electronic Ink

Electronic ink is another promising technology for screens on mobile devices. It is many times brighter than an LCD screen and draws an order of magnitude less power. The concept of electronic ink follows the logic that displays should look very much like print on paper and behave like print on paper. That means the display should be flexible like paper.

Electronic ink is processed into a film for integration into displays. Electronic ink is a fusion of chemistry, physics, and electronics to create this new material. Electronic ink is made up of millions of tiny capsules, about the diameter of a human hair, containing white particles suspended in a clear liquid. An electric field causes the particles to move to one side of the capsule or the other, creating either a dark spot or a white spot (Figure 8.24).

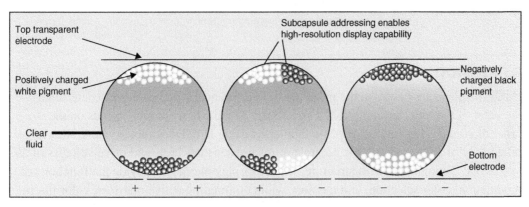

Figure 8.24: An Electronic Ink Display with Black and White Microcapsules Suspended in a Liquid

Source: http://www.elecdesign.com

An electronic ink display does not require a backlight and it is persistent. The persistent feature means that it does not require a constant supply of electric current to maintain a picture and is thus power friendly.

To form an E Ink electronic display, the ink is printed onto a sheet of plastic film that is laminated to a layer of circuitry. The circuitry forms a pattern of pixels that can then be controlled by a display driver. These microcapsules are suspended in a liquid "carrier medium" allowing them to be printed using existing screen printing processes onto virtually any surface, including glass, plastic, fabric, and even paper. Ultimately, electronic ink will permit most any surface to become a display, bringing information out of the confines of traditional devices and into the world around us (Figure 8.25).

Figure 8.25: Electronic Paper E Ink Corp. Display

Courtesy: http://www.boston.com

8.5.1.2 IMOD

IMOD displays are based on the principle of interference, which is used to determine the color of the reflected light. IMOD pixels are capable of switching speeds on the order of 10 μs. Additionally, displays fabricated using IMOD technology have shown reflectivity of greater than 60%, contrast ratios greater than 15:1, and drive voltages of as low as 5 V. Though simple in structure, IMOD display elements provide the functions of modulation, color selection, and memory while eliminating active matrices, color filters, and polarizers.

The IMOD pixel (see Figure 8.26) consists of a glass substrate which is coated with thin films. Beneath the glass is a reflective conductive membrane which is separated from the glass by a thin air gap. When a voltage is applied to the membrane and the thin films on the glass, the membrane experiences electrostatic attraction and is drawn toward the glass. This state is called the collapsed state and the pixel appears black as the light entering is shifted to the UV spectrum. The application of a lower voltage level returns the membrane to the original position called the open state. In this state the pixel appears bright and colored. This color is generated by interference of light, a process which is much more efficient than using color filters.

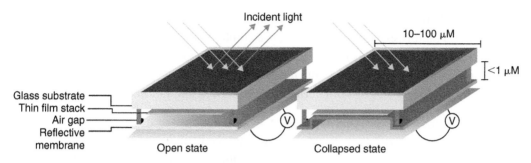

Figure 8.26: Basic Structure of an IMOD Pixel [16]

Source: http://www.qualcomm.com

The low power advantage of the IMOD display is due to the mechanical structure of the IMOD pixel. There is a hysteresis effect in the electro-optical function of the IMOD display. As a result, it requires very little power to maintain either one of its two states. This feature, called bistability, allows a display constructed of IMOD pixels to maintain an image with near-zero power consumption.

The IMOD element provides functionalities that would require three separate elements in an LCD, each of which require multiple processing steps. First, the modulation is accomplished via the movable membrane rather than the liquid crystal. Second, the color is generated using the air gap rather than color filters. Finally, the hysteresis provides the memory unit that a thin-film transistor (TFT) provides in an LCD.

8.6 Summary

Mobile devices have become so rich with features that they are almost more than can be utilized effectively. However, this trend shows no sign of letting up. New features are being introduced to keep consumers buying the latest models and developing the demand for applications and content. Applications such as games, Internet access, m-commerce and content similar to mobile TV, HD video, and 3D audio delivered at broadband speeds.

A barrage of new technologies, currently foreign to the mobile device industry, is required to meet the ever-growing demand of the mobile device consumer. Gordon Moore's law and its mapping into International Technology Roadmap of Semiconductors have been the most powerful drivers for the development of the microelectronic industry in order to converge with the consumer's hunger for mobile devices.

Nanotechnology involves the knowledge and manipulation of materials at the nanometer or the atomic scale to create structures that have novel properties and functions because of their size, shape, or composition. Although nanotechnology is still nascent, it is nevertheless a rapidly evolving field of science combining insights from physics, chemistry, biology, material science, and engineering. Nanotechnology combines existing knowledge and ongoing research at the nanometer level, with a specific emphasis on applying the science to engineer materials with enhanced and tailored properties.

Currently, nanotechnology applications are mainly in the form of relatively simple powders and coatings with various uses, as these are the main morphologies that current technologies allow synthesis on a large scale with uniform properties. Nanotechnology is extremely pervasive and expected to touch virtually all industries and end markets over the longer term. Indeed, nanoscale materials are already commercially used to enhance the properties of basic applications such as water repellant fabrics, antistatic mats, carbon fiber tennis rackets, and sunscreen. In addition, 0.1µm or 100 nm technology in the manufacture of IC chips is an example of evolutionary nanotechnology that already exists.

There is already a range of design options and prototypes for next generation nanoelectronic switching and amplification devices. The worldwide enterprise of engineering nanometer-scale electronic computers is well underway and growing. One can be confident that such a dramatically smaller computational engines will transform mobile devices and their need for high performance and low power consumption.

The objective of the silicon-based nuclear-spin field is a quantum computer with a scaleable architecture allowing large numbers of qubits, a high clock speed; the capability to perform quantum operations in parallel and a scalable manufacturing method.

Solid-state systems are the most promising candidates for meeting these demanding requirements. Kane published a breakthrough conceptual study for a silicon-based quantum computer that will meet many of these requirements [9].

Scaling up a solid-state computer to over a million qubits is a challenging goal. However, approximately 50 years ago, computer companies attempting to reduce the size of their machines were just becoming aware of a new strategy known as integrated circuits.

Research and Development centers and some commercial entities are actively working in the field of MEMS. Products like MEMS microphones, RF MEMS, wireless sensor networks, domestic robots, smart pills, and lab-on-chips are in various phases of commercialization. Forecasting their success is challenging due to the engineering challenges. However, the minute and low power micro-machined components being developed will benefit the mobile device industry. Commercial successes to date include

RF MEMS and MEMS microphones that in addition to the low power aspect also appeal to the thin form factor popular with mobile products.

Building a computer out of living cells sounds like science fiction, but the foundations of such a technology are already being laid down. Recombinant DNA technology is a powerful tool that already has a wide range of applications including therapeutics, engineered crops, farm animals, biomedical and molecular scale fabrication. Armed with this new tool, engineers are now trying to create new computer technology that consists of living cells.

Scientists have characterized and assembled a genetic component library. There has also been successful implementation of prototype circuit. With the appropriate tools, the engineer of bio-circuits can begin to design and produce large-scale circuits. The primary challenge is to develop the required tools to devise models and perform simulations that can accurately predict outcome of genetic networks.

Combining today's advanced nanoscale SoCs with the milliscale components that make up 90% of typical electronic systems would not produce the emerging digital convergence the world has been expecting. Innovative packaging techniques, like RCP and SoP, go well beyond Moore's law. They combine nanoscale SoCs with newly developed micro- to nanoscale, thin-film versions of discrete and other components. This technology embeds these components in a new type of package so small that it eventually will transform handhelds into multi- or mega-function systems.

Fuel cells are the most likely energy source to replace mobile device batteries in the near future. They have the potential to provide longer run times with less weight and volume than existing power solutions for a wide variety of portable electronic applications. However, there are a number of additional alternative energy sources worthy of note: a micro-gas turbine generator being developed at MIT Microsystems Technology Laboratory; a portable power source with 10–50 times the power density of state-of-the-art batteries. Others include micro-structures for vibration energy recovery, power harvesting shoes [14], environmental heat harvesting, and solar photovoltaic to name a few.

The future of Electronic Paper Displays is with organic electronics. Organic electronics is a branch of electronics that deals with conductive polymers, plastics, or small molecules. It is called organic electronics because the polymers and small molecules are carbon-based, like the molecules of living things. Properties such as bistability and power-off image retention make ePaper ideal for future mobile devices.

Power and energy management has grown into a multifaceted effort that brings together researchers from such diverse areas as physics, mechanical engineering, electrical engineering, design automation, logic and high-level synthesis, computer architecture, operating systems, compiler design, and application development. The book has examined how the power problem arises and how the problem has been addressed along multiple levels ranging from transistors to applications. We have also surveyed major commercial power management technologies and provided a glimpse into some emerging technologies.

In review, there are many target areas that the mobile device developer focuses on to provide optimized energy solutions to their customers:

- Having the right process technology. Materials science will do the heavy lifting.

- Creating advanced packages to minimize parasitics and reduce board area stay on track with Moore's law.

- Circuit techniques to enable low power designs.

- Component architecture and RF design that minimize power.

- Platforms that combine all these techniques into a cohesive mobile device.

- Software that controls the energy consumed by the hardware.

- EDA tools, modeling, and system-level simulation to enable the engineers to understand the power of their products and to identify areas to optimize performance and power.

- Creative and innovative scientists and engineers.

There is no silver bullet when it comes to energy efficiency. Power and energy optimization is a collaborative effort among all disciplines in commercializing a successful mobile device. By focusing on the entire mobile device and not just the components, a holistic approach to energy optimization, facilitates the mobile device industry in discovering the variables that can maximize energy efficiency, and continue to reap the performance gains predicted by Moore's law.

The field is still active, and that researchers are continually developing new algorithms and heuristics along each level as well as exploring how to integrate algorithms from multiple levels. Given the wide variety of micro-architectural and software techniques available today and the astoundingly large number of techniques that will be available in the future, it is highly likely that we will overcome the limits imposed by high power consumption and continue to build mobile devices offering greater levels of performance and versatility. However, only time will tell which approaches will ultimately succeed in solving the power problem.

References

[1] B. Vigna. More than Moore: Micro-machined products enable new applications and open new markets. *Electron Devices Meeting, 2005. IEDM Technical Digest. IEEE International*, 5–7 December 2005, 8 pp.

[2] S. Mize. Toward nanomaterials by design: A rational approach for reaping benefits in the short and long term. White paper, http://www.scottmize.com, September 2004, pp. 5–9.

[3] Sematech, http://ismi.sematech.org/meetings/archives/other/7917/3_FEP.pdf, 15 March 2006.

[4] K. L. Wang. *Nanoarchitectronocs and Nanoelectronics. 8th International Conference on Solid-State and Integrated Circuit Technology, 2006. ICSICT '06*, October 2006, p. 8.

[5] R. Chau. Plenary talk, *14th Biennial Conference on Insulating Films on Semiconductors 2005 Leuven*, Belgium, 22–24 June 2005.

[6] G. Bourianoff. The future of nanocomputing. *Computer*, Vol. 36, No. 8, 44–53, 2003.

[7] P. G. Collins and P. Avouris. Nanotubes for electronics. *Scientific American*, December 2000.

[8] G. Bourianoff and ITRS 2001. The future of nanocomputing presentation, http://www.lems.brown.edu/~iris/en291a10-05/Lectures/DestaNanofuture.pdf. September 2005, pp. 1–45 of destananofuture.pdf.

[9] B.E. Kane. A silicon-based nuclear spin quantum computer. *Nature*, Vol. 393, 133, 1998.

[10] D. Baker, G. Church, J. Collins, D. Endy, J. Jacobson, J. Keasling, P. Modrich, C. Smolke, and R. Weiss. Engineering life: Building a fab for biology. *Scientific American*, Vol. 294, No. 6, pp. 44–51, 2006.

[11] R. Weiss, S. Basu, S. Hooshangi, A. Kalmbach, D. Karig, R. Mehreja, and I. Netravali. *Genetic Circuit Building Blocks for Cellular Computation, Communications, and Signal Processing*. Netherlands: Kluwer Academic Publishers, 2002.

[12] B. Keser. Birds-of-a-Feather: Redistributed Chip Package (RCP) Broad-Range Applications for an Innovative Package Technology. *Freescale Technology Forum*, June 2007.

[13] R. C. Phal and J. Adams. *Systems in Package Technology (Presentation), International Electronics Manufacturing Initiative*, SIP TIG report, June 2005.

[14] J. A. Paradiso and T. Starner. Energy scavenging for mobile and wireless electronics. *Pervasive Computing, IEEE*, Vol. 4, No. 1, 18–27, 2005.

[15] R. Allen. Flexible Displays set to go Mainstream. Electronic Design Online ID #15470. May 2007.

[16] L. D. Paulsson and L. Garber. News Briefs. *Computer*, Vol. 39, No. 10, 19–21, 2006.

Index

Printed and bound by CPI Group (UK) Ltd, Croydon, CR0 4YY

03/10/2024

01040334-0001